11.95
A

Systems Engineering Methods

Wiley Series on Systems Engineering and Analysis
HAROLD CHESTNUT, Editor

Chestnut
Systems Engineering Tools

Wilson and Wilson
Information, Computers, and System Design

Hahn and Shapiro
Statistical Models in Engineering

Chestnut
Systems Engineering Methods

Systems Engineering Methods

Harold Chestnut
Information Science Laboratory
Research and Development Center
General Electric Company

John Wiley & Sons, Inc. New York London Sydney

Copyright © 1967 by John Wiley & Sons, Inc. All Rights Reserved. This book or any part thereof must not be reproduced in any form without the written permission of the publisher.

Library of Congress Catalog Card Number: 67-17336
Printed in the United States of America

SYSTEMS ENGINEERING AND ANALYSIS SERIES

In a society which is producing more people, more materials, more things, and more information than ever before, systems engineering is indispensable in meeting the challenge of complexity. This series of books is an attempt to bring together in a complementary as well as unified fashion the many specialties of the subject, such as modeling and simulation, computing, control, probability and statistics, optimization, reliability, and economics, and to emphasize the interrelationship between them.

The aim is to make the series as comprehensive as possible without dwelling on the myriad details of each specialty and at the same time to provide a broad basic framework on which to build these details. The design of these books will be fundamental in nature to meet the needs of students and engineers and to insure they remain of lasting interest and importance.

Preface

In a world in which the training and the functions of individuals and groups are growing more and more specialized the number of ways to accomplish any particular result is increasing. Different designs, different facilities, different equipments, different methods, and different organizational means are available to meet the needs of man. It is highly desirable that we have trained persons to look at these varied possibilities, to compare their effectiveness, and to point the way to sound engineering decisions. *Systems Engineering Methods* is directed toward the development of a broad systems engineering approach to help such people improve their decision-making capability. Although the emphasis is on engineering, the systems approach also has validity for many other areas in which the emphasis may be social, economic, or political.

Each system is unique. However, by capitalizing on the similarities between systems we can reduce the time, effort, and cost for some or all of them, and thus the quest for formalized design methods becomes more attractive.

This book represents a change in emphasis from the ideas presented in my earlier book *Systems Engineering Tools*. The first book stressed the different technical and analytical tools used in systems engineering, whereas this book is concerned with the problems of the systems engineer who has the over-all responsibility for the success of a system.

I have made an effort in Chapter 1 to introduce sequentially the various functions and evaluations that are required before a system can be engineered. Chapter 2 points up the need for recognizing all the steps to be performed in realizing an operating system and for organizing systematic ways to perform them. The importance of formulating and structuring a system and the various methods by which it can be achieved are described in Chapter 3.

System evaluation is crucial to decision-making or optimization in any system. Several methods for establishing a system's value are presented in Chapter 4. Chapter 5 discusses classical bases for arriving at system costs and cost-benefit and cost-sensitivity considerations. Chapter 6 focuses on various meanings of the word "time," which is one of the factors common to the different parts of a system. Control of system time by scheduling, PERT, CPM, and Task Network Scheduling are given particular attention. Reliability from system concept to system operation is treated in Chapter 7. In Chapter 8 the major systems engineering methods and tools are related, and many areas in which systems engineering will receive special attention in the future are highlighted. Although the order in which these chapters are presented is logical, the topics can be covered in a different sequence without reducing their value.

My experience as a member of the General Electric Research and Development Center has had strong influence on the material contained in this book. However, the ideas presented here are not necessarily those of that company. They reflect the influence of many individuals and organizations with whom I have been associated, such as the IEEE, the American Automatic Control Council, and the International Federation of Automatic Control. Perhaps it is my awareness of the many ways in which things are done by these people that has pointed up the need for and influence of systems engineering.

This book has been greatly improved by the help of many people, among whom are Wallace Barnes, Ernest Bianco, Dudley Chambers, Rowe Chapman, John Coales, Charles Concordia, A. D. Hall, R. Harvey, E. Iozzino, F. V. Johnson, L. K. Kirchmayer, Otto Klima, D. F. Langenwalter, Irving Lefkowitz, Victor Louden, John Lozier, R. N. Mayer, Mihailo Mesarovic, W. E. Miller, R. Hosmer Norris, Vincent Picozzi, G. Reethof, and A. Yerman. The help of Mary Ferrucci in typing the text and handling the manuscript has been invaluable. Lastly, I am grateful to my family for their patience and encouragement.

Schenectady, New York   HAROLD CHESTNUT
March 1967

Contents

1 The Environment for System Engineering Methods

1.0 Introduction	1
1.1 Establish the Value or Need for the System	9
1.2 Determine the System Cost	12
1.3 Estimate the Time Required to Produce the System	15
1.4 Formulate and Structure the System More Specifically	18
1.5 Organize and Outline the Effort to Do the Job	20
1.6 Perform the Work Necessary to Ensure a Reliable System	21
1.7 Conclusions	22

2 System Organization, Scheduling, and Record-Keeping

2.0 Introduction	24
2.1 System Engineering Plan of Organization	25
2.2 Scheduling and Review	41
2.3 Subsystem, Equipment, Assembly, and Component Identification	44
2.4 Record-Keeping and Communications	52
2.5 Check-out and Servicing	54
2.6 Similarities and Differences	61
2.7 Specifications	62
2.8 Design Changes	65
2.9 Conclusions	69

3 Formulating and Structuring the System

3.0 Introduction	70
3.1 Problem Objectives and Goals	72
3.2 System Environment	75
3.3 What Are the System Inputs?	87
3.4 What Are the System Outputs?	92
3.5 Problem Structuring and Its Relation to Problem Formulation	98
3.6 Functional Structuring	107
3.7 Equipment Structuring	121
3.8 Conclusions	133

4 Factors for Judging the Value of a System

4.0 Introduction	135
4.1 System Requirements Selected Strongly Influence the Value of a System	136

4.2	Performance as a Measure of Value	145
4.3	Net Return Basis for System Evaluation	155
4.4	Value of a System	161
4.5	Effect of System Time Phase on Criteria for Judgment	171
4.6	System Value as a Function of Performance, Cost, Time, and Reliability	173
4.7	Conclusions	178

5 Cost

5.0	Introduction	179
5.1	What is the Product or Service Being Costed?	180
5.2	Influence of Other Factors on Cost	184
5.3	Long-, Medium-, and Short-Term Cost Considerations	187
5.4	Build-up of Total Cost from Cost of Parts	190
5.5	System Costs Considering Only Variable Costs	201
5.6	Cost-Benefit Analysis	210
5.7	Conclusions	219

6 Time

6.0	Introduction	221
6.1	Different Meanings of Time	222
6.2	Effect of Time on Other Factors	227
6.3	The Two-Time Boundaries—Now or Then	233
6.4	Time to Make System versus Time Required for System Itself to Operate	237
6.5	Time Schedules	241
6.6	PERT	243
6.7	Technicalities of Network Analysis	247
6.8	Task Network Scheduling	255
6.9	Conclusions	268

7 Reliability

7.0	Introduction	269
7.1	Reliability and What Is Being Done About It	270
7.2	Reliability Arithmetic	277
7.3	Designing for Reliability	284
7.4	Controlling Quality for Reliability	336
7.5	Maintenance for Reliability	346
7.6	Management with Reliability in Mind	349
7.7	Conclusions	354

8 Conclusion and Prologue

8.0	Introduction	355
8.1	Summary of Tools and Methods	357
8.2	Typical Systems Opportunities for the Future	361
8.3	Conclusion	378

Bibliography	379
Index	383

Systems Engineering Methods

1

The Environment for System Engineering Methods

1.0 Introduction

The concept of a system is not a simple or unique one. There are many different kinds of systems, and different systems may be organized and operated in different ways. As individuals we all belong to some social system, we participate in an economic system, we are the product of several educational systems, and we are members of one or more family systems. In a similar fashion, the equipment of which physical systems are made may be members of many other systems, such as electrical, mechanical, sensing, actuating, energy, materials, and/or information systems. One of the challenges to the person who engineers a system is to find the many alternative ways in which the function, the operation, and/or the equipment of concern and interest may be considered, understood, and made to perform most effectively.[48]

Included in the problems of systems engineering are those of complexity and of choice. Of all the available facts about a system or the needs for a system, which are of most significance for the present circumstances and for their probable future course? How much information is needed and how should it be used to make a satisfactory decision, considering the time and resources available and the purpose to which these data are to be applied? Since most of the means of understanding which we as individuals use, or which are used by the automatic decision-making processes which we employ, are serial processes, we are continually faced with choices of how to divide the jobs to be done and to select an order or an arrangement for systematically handling the abundance of data which are available. Alternatively, there may be too little of a resource—time, money, materials, data, or manpower as examples; how can these be systematically handled to achieve their most favorable utilization?[32]

In this book on systems engineering methods, the emphasis is placed on the problems associated with organizing and performing the jobs that are necessary to accomplish the process of making the system itself. In order to have some objective means of evaluating which of the alternative methods for producing the system is most appropriate in a given case, it is essential to understand the many judgment factors involved, such as the value of the system being planned, its cost, the time required to produce it, and its reliability. Therefore, this book on systems engineering methods is devoted principally to the higher-level decision-making processes of systems engineering which help to establish the goals and the trade-offs between conflicting objectives with which systems judgments must contend.[46, 47, 61]

Systems Engineering Methods builds on the information contained in the companion volume, *Systems Engineering Tools*,[49] which preceded it. The earlier book emphasized the necessary engineering, mathematics, and judgment principles involved in performing many of the detailed portions of the systems engineering job. Included in these systems engineering tools are modeling and simulation, computing, control, probability and statistics, signals and noise, optimization, and error analysis. The stress was on analysis and synthesis methods in general, and the result has been to provide a sound basis for performing effectively at any one of several levels on a systems job. The analysis has tended to be devoted to the problems associated with how the operating system performs. Hence, the tools have endeavored to be very broadly applicable and to be useful to anyone contributing to a part of a system or the major system itself.

However, the same tools can be effective when applied to problems associated with the organization and performance of the work functions or jobs necessary to accomplish the process of making the system itself.[26] Therefore, many of the tools developed in the preceding volume will find use in our consideration of the systems engineering methods described in this book.

The more fundamental questions—What is the worth of a particular system? Should it be built at all? How should it be structured and organized? What are reasonable values of costs and time for producing it?—these are definitely higher-level decision-making topics of systems engineering which a much smaller number of people in responsible positions of leadership or management will be called upon to make. Because these matters of broad systems engineering judgment are generally not ones that novices in the field need cope with early in their career, they were not stressed in the earlier volume. The fact that these topics were not presented until this second book

Introduction

by no means indicates that they are subordinate in importance or that they should occur second in time as part of the systems engineering process. Their presentation was delayed because fewer people need this information initially to perform useful work on a project and because more than a little of this important decision-making process has a high degree of subjectiveness and the "art" of individuality about it. It tends to be creative in nature and not so easily amenable to firm rules for methods of performing it. Although many individuals working on systems may seldom if ever be called upon to make top-level systems engineering decisions, as many people as possible should understand the systems engineering approach at higher levels since these decisions provide a part of the environment within which lower-level decisions take place and with which they must be compatible.

Definition of Systems Engineering

In *Systems Engineering Tools*[49] the definition of systems engineering was set forth as follows: *"The Systems Engineering method recognizes each system as an integrated whole even though composed of diverse, specialized structures and subfunctions. It further recognizes that any system has a number of objectives and that the balance between to optimize the overall system functions according to the weighted objectives and to achieve maximum compatibility of its parts."*[20]

If we think of the overall problem of systems engineering as being composed of two parts, one being the system engineering associated with the way that the operating system itself works and the other the systematic process of performing the engineering and the associated work in producing the operating system (see Figure 1.0-1), it is apparent that the two parts may strongly influence each other. In this book the problems to be emphasized are those associated with such processes as engineering, manufacturing, shipping, and installing and maintaining the operating system. Figure 1.0-2 shows in more detailed block diagram form the different parts of each of the two aspects of the overall systems engineering problem. It is inevitable

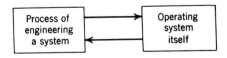

Figure 1.0-1. Interaction of process of engineering a system and the operating system itself to make up the broader systems engineering activity.

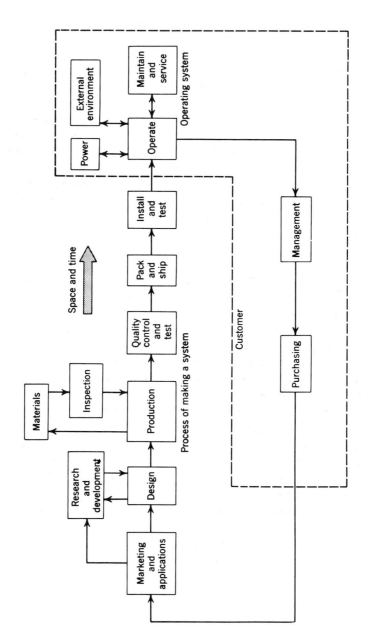

Figure 1.0-2. Interrelationship of process of making a system and the operating system itself.

4

Introduction 5

that the influence of the characteristics of the operating system be felt by the process of making the system. A number of these interaction influences will be considered in this book.

Defining a Need for the System

Characteristically, systems engineering tends to be associated with poorly defined problems in that many times at the outset the customer of the system may not know exactly what the system should be or do. It is frequently necessary to help him understand what is realizable, in what time and at what cost. Since it may not be clear initially exactly what the system should do, or the cost of the desired system may exceed the money available to buy it, it is difficult to describe just what system performance can be obtained and what cost and time will be required.

However, since this system will have certain similarities to prior or currently existing systems and will have to compete with alternative existing or proposed methods, the field of choices available tends to be somewhat limited. Furthermore, the work required to achieve this system will tend to be similar in many ways to the generic work elements that have been used successfully on existing systems. Included in the elements of work to be done to realize a system are those shown in Table 1.0-1.

Although the above steps are listed in an order, the sequence in which they are performed may not always be strictly as shown here, nor is the order always the same for all systems. As pointed out in the first chapter *Systems Engineering Tools*[49] and described later in Section 1.4 of this chapter, the process is an iterative one in which considerable ingenuity and vision are needed to piece together the relationships which must exist for a successful solution. The remark "You can't get there from here"—that is, the system as proposed cannot accomplish the desired objectives—may be correct for the route (solution method) that is being considered, but this merely means that another one or ones must be tried which hopefully can more nearly meet the objectives being sought.

Table 1.0-1. Generic System Work Elements

Establish the value or need for the system.
Determine the system cost.
Estimate the time required to produce the system.
Formulate and structure the system more specifically.
Organize and outline the effort to do the job.
Perform the work necessary to ensure a reliable system.

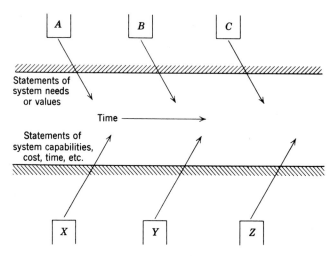

Figure 1.0-3. Interaction of systems users and systems suppliers with time to derive new systems.

Referring to Figure 1.0-3, one notes that there tends to be a continuing interaction with time of systems users and systems suppliers. Thus, for example, systems *A, B, C,* which may represent different companies, different governmental organizations, or different individuals within each of these or other groups, are, as a function of time, gaining experience and/or expressing interest in using various systems. Persons in operations research and development or planning are typical sources of such potential needs. From these experiences and interests are generated statements about implied or actual needs for systems or functions to be performed.

Also as a function of time, in the category of system suppliers, such as *X, Y,* and *Z,* people are gaining experience and developing interests in supplying various sorts of systems for possible users. Persons in marketing, applications, and research and development are either charged with the responsibility for generating various new systems for which estimates of performance, cost, time to produce, etc., are developed, or else spontaneously produce such innovations.

Actually, the individuals or groups in the suppliers and/or users categories are not always well defined or clearly established. Because of time seniority, technological change, or relative competitive pressures, the members of these groups tend continually to undergo change.

Figure 1.0-4 shows one possible way in which a system user and a system supplier may interact. On the basis of his exploration of

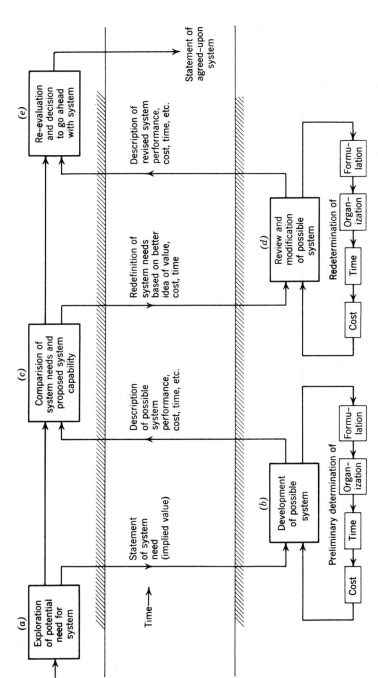

Figure 1.0-4. Interaction of system users and system suppliers to establish system need and value.

the potential need for an existing or hypothetical system (*a*), the user produces a statement of a need for a system with an implied value for it. This information is used by the supplier as a basis for a tentative development of one or more possible systems capable of meeting in whole (or in part) the user's needs (*b*). To do this, the supplier has to formulate and structure the problem to some degree, estimate an organization or method for accomplishing the work, determine how long it will take to make the system, and perform an estimate of the cost required to produce it and put it in operating order.

From this effort on the part of the supplier comes a description of a possible system in which its performance, cost, time to realize it, and other pertinent characteristics are defined. Provided with this more detailed information, the user can now perform a comparison of the system needs and the capability of the proposed system (*c*). The result can be a redefinition of the systems needs based upon a better idea of value, cost, time, flexibility, etc.

At this point it is of interest to note that the exploration and comparison functions (*a*) and (*c*) performed by the system user may in fact be carried out by individuals within the supplier's organization who are charged with these responsibilities. Furthermore, the user may likewise decide that the system development function (*b*) should be done by people in his own organization. In these ways, organizations find that their roles can change with time as they develop or acquire people with additional skills, capabilities, and experience and as they obtain facilities and equipment.

From the system redefinition (*c*) may come the need for review and modification of the possible system (*d*), an iteration and description of the revised system and its performance, cost, time, and other characteristics. Presumably a re-evaluation by the user and a decision to go ahead with the system (*e*) result in a more definitive statement of the system agreed upon—its characteristics, its cost, its trade-offs or weighting functions—and a firmer basis upon which work can proceed.

In this chapter, the various elements of work listed in Table 1.0-1 will be discussed in their broad aspects. Although these work functions are ones that may be done by different groups of individuals at different times and perhaps for different purposes, it is well to view them briefly in their relationship to one another. The details of each can be time consuming and expensive to achieve. Hence, each tends to be handled somewhat independently of the other, although in fact it is highly desirable that they be closely integrated.

Partitioning of the jobs or work functions in this fashion can serve a useful purpose with respect to simplifying the information supplied to each group, but the lack of overall knowledge or integration may allow errors or misunderstanding to occur in various places.

1.1 Establish the Value or Need for the System

Systems of the sort considered in this book generally involve a significant amount of resources or cost to the organization which is obtaining them.[64] Typical of such resources requirements are money, manpower, equipment, material, or a combination of several of these. Hence it is necessary to obtain management or corporate approval for the use of these resources. Without such approval, it is very difficult to bring into being a system of some size or significance.

In order to obtain management approval for the allocation of the required resources, it is essential that the worth or value of the system relative to its cost be sufficient to warrant the expenditure. It is also essential to establish the technical feasibility of the undertaking, but this is generally done initially on a broad basis rather than in minute detail.

As shown in Figure 1.1-1, the organizational hierarchy for the approval of systems of increasing size or cost involves the higher levels of corporate management. To obtain this higher-level corporate or organizational management interest and support of a system it is necessary that the values for the system be expressed in terms of objectives which are meaningful to people in such positions of responsibility. Return on investment, profit ratios, product leadership, technical innovation, prestige, security for the present or future life of the organization are typical of these essential values. Although it will be necessary in the earlier stages of gaining acceptance for the

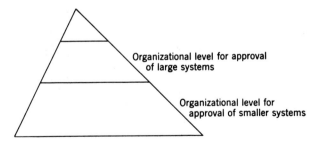

Figure 1.1-1. Organizational hierarchy for approval of systems of various sizes.

feasibility of the proposed system to justify its technical soundness, at the time when initial formal approval is needed to go ahead with the system, the technical features will generally not be the paramount considerations.

It is of interest to note that the initial life or death, that is, go or no-go, decision with respect to a proposed large-scale system is generally made by a relatively few people and on bases which are essentially nontechnical. Since many years may elapse between initiation and completion of a system, precise technical data may be impossible to obtain. Therefore, it is important for the principal engineer, in addition to trying to ensure the technical soundness of his proposed system, to provide an understandable and attractive set of values and/or value/cost considerations with which to impress the decision makers. Occasionally systems engineers are so concerned with the technical elegance of their proposed project that they fail to realize the importance of a well-balanced set of value objectives for their system. As a result they may not be able to convince management and to gain the necessary approval for their system.

The value objectives referred to here may be represented typically by such well-accepted criteria of overall value or performance index as technical performance, cost, time, reliability, and flexibility, appropriately weighted as shown in Figure 1.1-2. Thus, the overall system value may not be a single-valued scalar function, nor need it necessarily be an additive function as shown in Figure 1.1-2. Perhaps, the ratio of performance to cost shown as a function of complexity in Figure 1.1-3 may be employed as a criterion of value. Other criteria

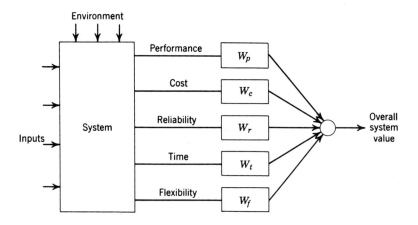

Figure 1.1-2. Overall system value as an additive weighting function of specific system outputs.

Establish the Value or Need for the System

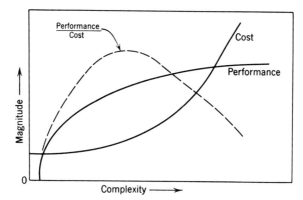

Figure 1.1-3. System performance/cost as a function of complexity as a possible criterion of value.

of performance, such as efficiency, speed of response, reaction rates, and other figures of merit, may likewise be introduced to provide other bases for the judgment of system value.

Consider now the environment in which the value of a proposed system must be established. In keeping with the concept that there are alternative ways of accomplishing a given objective, a proposed system must in effect compete with some existing or other proposed methods for accomplishing a similar or relatively comparable result. This means that as a point of reference there are alternatives available to the person or organization which might require that the system be built.

As an example, the proposed system may have for one or more of its value objectives the fact that its performance in some way or ways must exceed that of existing competitive systems or other alternative methods. On the other hand, the cost of the new system may have to be less than that of an existing system. The time required to obtain the proposed system may be shorter, or its life may have to be longer, or it may perform its task more quickly. There are many aspects of time and its use which would influence the selection of the system. Furthermore, the reliability of the proposed system may have to represent a significant improvement over that of the existing system or other competitors. Thus, before significant sums of money, manpower, and material can be devoted to engineering and building a system, it is necessary to establish that the system as proposed will perform a needed function or combination of functions with values such that it can be justified in comparison to the alternative ways or existing methods.

Although from a technical point of view a proposed system may be difficult or challenging, of itself such a factor is not necessarily sufficient to justify the choice of a new or proposed system. Whereas it may be necessary to show technically that a particular scheme or method is feasible before proceeding with adoption or approval of a system, it is essential that some person or persons be convinced that the proposed system is needed and has sufficient value to warrant its being constructed.

In the case of military or space systems the nature of the decision factors and value judgments[50] that are involved may be quite different from the corresponding factors for industrial, utility, or commercial systems. The nature of the threat to the nation's security may be such as to place a very high value on a proposed new military system even without complete details of the system or even before its full formulation has been started. Similarly, the importance to the purchaser of a scientific or a prestige type of undertaking, such as a manned trip to the moon or to Mars, may be sufficient in itself to justify a high value for the system.

In the case of an industrial system, for example, an airline reservation system or an improved hot-strip steel mill, the proposal can be evaluated in fairly concrete financial terms, and existing and/or alternative methods should receive very close scrutiny before any significant amount of detailed systems engineering effort can be justified.[60] Generally, experience from previous or similar systems is drawn upon heavily to provide estimates of formulation, organization, performance, cost and time, so that fairly definite benchmarks may be available for both value and cost.

Of course, the value need not necessarily be economic; for such reasons as strategy or prestige a system can have significant if intangible value. If so, it should be understood that these other reasons represent the basis for the system value.

The value of the system is stressed at this point because this factor may be of overriding importance in establishing whether or not the technical effort required to make a system work is justified or can be undertaken. Value, in any one of the several senses described above, rather than technical achievement, is frequently a major basis for the decision to proceed with a system.

1.2 Determine the System Cost

Another important factor in extablishing the overall worth of a system is the magnitude of its cost. Thus, once the system value

Determine the System Cost

has been established, it is necessary to determine the system cost as a basis for comparison.

One of the few things that can be considered to be known with reasonable assurance by the purchaser of a system is that its cost will not be less than that quoted. This is especially true of a fixed-price job in which the vendor agrees to supply a given equipment or service for a specified sum of money at a certain time. However, sometime costs are quoted on a cost plus fixed-fee basis, in which event no fixed price is associated with the system and the purchaser may thereby be required to pay more than the initial estimate of the cost. Frequently, the cost is quite difficult for the maker of a system to determine because it depends not only on the particular system being designed and built but also on what other jobs may be worked on at the same time. These other systems may be able or required to share the facilities, manpower, and skills necessary to accomplish the job.

System costs can be considered in many different ways. For example, the cost for making the system and the cost for operating it may be used as a basis for judging the total cost. The total cost may also be arrived at in another fashion, as, for example, the sum of the fixed cost, variable cost, past investment and other charges.[31] Furthermore, in many cases where systems engineering is required several time phases are involved for which the cost factors may be significant, such as the study phase, the breadboard phase, the prototype phase, and the production phase. In the study phase the principal emphasis is on theoretical considerations. In the breadboard phase experimental methods are employed to demonstrate the significant parts of the system so that some equipment costs will be included. In the prototype phase actual hardware and equipment are employed so that increased equipment costs result. The production phase requires complete tooling, maintenance, etc., in addition to production and other costs. Obviously the cost associated with each of these phases will be influenced not only by what must be accomplished in the particular phase but also by what is, has been, or will be required of the associated phases of the same system. Also, depending on the production quantity involved, the extent of tooling, the amount of installation and maintenance costs, and the associated costs of training and replacement parts may contribute significant cost factors. Table 1.2-1 shows one form of total cost breakdown applicable to military systems. Interestingly enough, in some cases the related costs may be as large as or larger than the cost of acquisition and ownership.

The result of the cost determination may have an appreciable effect

Table 1.2-1. A Form of Total Cost Breakdown

I. *Acquisition and Ownership of Article Itself*
    Development and test
    Production
    Operation and maintenance

II. *Related Costs*
    Data
    Personnel training and training of support facilities
    Maintenance and maintenance support facilities
    Transportation and storage
    Integration
    Assembly and check-out
    Spares
    Logistic support facilities

on the decision as to whether or not the value of the system sufficiently exceeds the cost to justify going ahead. The term *cost effectiveness*[52, 56] is used in this regard to compare one system with another. As such, the cost figures are useful to help establish the answer to the question, "Does the system or equipment being proposed have sufficient utility to warrant the expenditure of resources that will be required to develop and use it?"

Another way of breaking down the total costs is into the categories of fixed, variable, and past investment and other charges. Fixed costs are dependent strongly on the magnitude of the initial capital investment in the system and are relatively insensitive to the method of operating the equipment. The variable costs are charges which are dependent on the method of operating the system but are also influenced by the characteristics of the system as well as the level of the output or yield. The past investment and other charges are costs dependent on factors that are essentially outside the control of the manager of the system but nevertheless alter the charges for the facility or process.

The subject of cost may be approached from two different points of view. In the first, the system requirements and specifications must be met completely, and the cost is a related quantity which is established by the specifications and is quite dependent on them. The second approach is that the cost for the system is relatively fixed, and the extent of the system performance and its characteristics are to be determined within the general cost level established. Thus, if a given system cost is established a priori, the performance and value of the system which will result become dependent conditions and are to be optimized in the framework of the customer's and the system's needs.

The thought that cost concerns only the financial and marketing people but has no technical and engineering importance is an unfortunate misconception. Costs help establish a basis by which value is judged and to which the performance and other characteristics which can be obtained from a system are inexorably linked. When systems costs have been prepared to include the necessary engineering and technical studies and investigations to determine stability, performance, manufacturability, and reliability, studies of these factors can be made with a higher likelihood of systems success than if such costs have been omitted. Cost constraints can have a significant bearing on the technical accomplishments possible for a system.

In recent years, the cost of the preliminary systems engineering studies has been recognized more realistically as a significant and legitimate item of expense. Frequently, study contracts have been let in the fashion of consulting efforts in which one or more contractors are paid for an investigation of the technical aspects of a proposed system. From such problem definition studies[43] a more accurate estimate of the technical, time, and cost requirements of the system is obtained and more realistic cost/effectiveness evaluation can be made. The price of many proposed systems is so high that cost and feasibility estimates obtained from the supplier at no direct expense to the purchaser may be more costly in the long run than ones which are paid for separately. System suppliers frequently are willing to invest some of their own money in performing on study contracts but are unable to justify supplying the high level of technical competence required to perform complete system studies for no direct compensation.

Cost has a very important influence on the judgment given a system during the decision-making process as well as on the technical achievements which the system can ultimately realize. It should be an important consideration in systems engineering.

1.3 Estimate the Time Required to Produce the System

The value of the system and the system costs are important factors. So also is the time required to produce the system. In reality, of course, the time required for and value of the system are somewhat dependent on the time available to do the systems job, and the cost of the system has been based on the assumption that the system can be built and realized in some reasonable period of time.

Here, as in many semantic problems, the word "time" is used by different people to mean different things. Such meanings as *total elapsed time* (calendar time from start to finish of a project), *time required to perform the actual task* (the number of hours of service to be paid for), as well as many other expressions involving time are frequently referred to loosely merely as time. Hence, it is desirable to be specific in talking about the time required to produce the system. Since systems costs may involve large sums of money and/or large numbers of people, additional time taken by one part of a system, even at no direct expense to that part, may result in large charges being incurred by the overall project because of constant expenditure rates that occur in other areas. The expression "time is money" is appropriate here, and proper attention in a systems sense must be given to the time it takes to accomplish most projects.

Two convenient methods are available for estimating the time required for a project. One procedure is to start with the time at the beginning of a project and to proceed to estimate in a forward sense by summing up the time required for each part of the total job to be done. The other way is to start from the time that one is required to have the system available and work backward through the process of obtaining the system in a reverse time sense so as to move from the completion date back toward the present. Figure 1.3-1 illustrates these two methods. Note that jobs A-A', B-B', C-C', etc., may be exactly the same under the two different methods. However, the elapsed time for each part may be different because of varying manpower or other allocations, and the costs and risks may be different as well. Each of these methods of estimating time has its advantages and drawbacks, and there is merit in trying to use the best of each.

From a time control viewpoint, it is necessary and desirable to put together time schedules which endeavor to describe the future in a clearly understandable fashion, so that all the parties involved can understand their roles and time requirements as well as the roles and time requirements of the others associated with the system.[26, 35]

It is important in performing the time estimation to realize what alternatives are available in reaching certain time objectives. For example, if more manpower is placed on the job, it will in general be possible to shorten the time required to produce the system. Conversely, if less manpower is available, this factor may stretch out the time before the system becomes available.

Estimate the Time Required to Produce the System

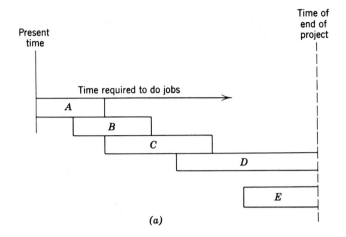

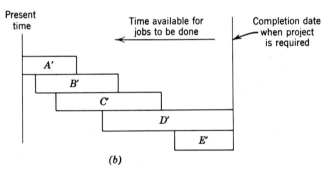

Figure 1.3-1. Comparison of computations of time for accomplishing project. (a) Starting at present and working forward. (b) Starting from completion date when projects are required and working back.

In similar fashion a shorter time for obtaining the system may require additional cost, for example, as overtime payments. Also, if a system is obtained in a shorter time, certain sacrifices in performance or reliability may result. Thus, the estimate of the time required to produce the system is influenced by assumed performance, reliability, cost, and other factors which in turn will be modified depending on the time selected as that in which the system should be produced. Task Network Scheduling (TANES) and PERT (Program Evaluation Review Technique)[14] are two methods used in trying to control the time it will take to realize a system.

1.4 Formulate and Structure the System More Specifically

Up until this point the system that has been considered from the point of view of its value, and for which the cost has been obtained and an estimate of the time required to produce it has been made, has in reality been merely a proposed system. Actually, the job of designing a system in detail is an expensive activity. A person or organization can, in general, afford to do the design job in detail only when being paid sufficiently for this work. In many cases, when the value, the cost, and the time are being determined, insufficient funds have been allotted to justify detailed formulation and structuring of what is merely a possible or proposed system.

Only when a system has definitely been authorized, or perhaps when a specific contract has been let for a detailed design study, can the specific formulation and structuring of the system in detail be justified and performed. Formulating the system consists in determining its inputs, outputs, requirements, objectives, and constraints. Structuring the system provides one or more methods of organizing the solution to the method of operation, the selection of parts, and the nature of their performance requirements.

The place of the formulating and structuring of the system[49] is indicated in Figure 1.4-1, in which they are shown to be an intimate part of a loop which also includes external information about the problem, feedback information about the system, and feedback information about the problem. Thus, more specifically the formulating and structuring of the system consist of a detailed, technical appraisal of the specific needs of the system. This appraisal is much more strongly oriented in terms of what is needed for the system to accomplish its objective than in terms of the more customer-oriented considerations of value, cost, and time which have previously been discussed.

From the formulating and structuring of the system, detailed specifications arise which provide information about the inputs and outputs and subsystem requirements. In addition, the structure provides a way of having information about one part of the system available to provide inputs to parts subsequently related, either geographically, physically, or perhaps in time. This part of the techniques problem is a highly creative one and is strongly dependent on the state of the art in which the system is to be built and operated. It also must include considerations of what will be necessary to support, that is, train, service, and maintain, the system.

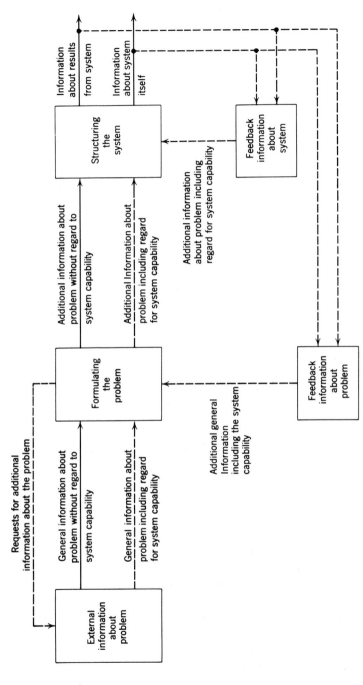

Figure 1.4-1. Information flow diagram showing relationships between formulating, structuring, and other aspects of the systems problem.

1.5 Organize and Outline the Effort to Do the Job

Once a definite idea has been established as to what the system is to be and what its specifications are, it is possible to formulate an organization and pinpoint the efforts required to accomplish the job.[36]

The organization must strike a balance between program- or project-oriented activities, which are related to the particular system and its needs, and technology-oriented activities, which may have continuity over many systems and are skills basic to the accomplishment of any system or program. Thus, it is possible for technology-oriented activities to function through many projects and to contribute to them. There must also be designed an evaluation-oriented activity as well as broad concept-oriented activities which provide new ideas for other systems in the future. Thus the basic organization for systems effort has many specific objectives in itself.

The effort to do a systems engineering job involves several different factors. One is an identification of the overall systems requirements and their criteria. This corresponds roughly to the formulation and structuring of the system mentioned in Section 1.4.

The next is the preparation and performance of the operational system. Presumably this consists of many particular jobs, but these are organized with the objective of making the system itself work. Another area in which effort must be taken is the performance of the necessary functional analysis and systems thinking, which involves supplying the support systems which back up the operational system. Another sort of effort which is essential is a reconciliation of the needs of the operational and the support systems and a comparison of the proposed design with the required overall systems requirements initially set forth.

Another aspect of the organization and the outlining of the effort to realize a project consists of setting forth scheduling and review procedures and establishing methods of identifying items of equipment in terms of drawings, name plate data, and other means of locating specific terminal boards and other positions.

Associated with the identification of equipment is the need for a record-keeping procedure to maintain up to date the information about the system as it becomes available and is required by people later on in the life cycle of the system, for example, when it is being procured and built. Other areas where organization is needed are in the check-out means for the system, in the specifications and the test of the system, and in providing means for permitting the system to grow and change, as through alterations.

All these requirements for organization are ones which should be established initially. Systematic and clearly stated ways for accomplishing them can perform a useful and worthwhile function in terms of aiding in the communication to the people and organizations which are responsible for the development, production, installation, and use of the actual system.

1.6 Perform the Work Necessary to Ensure a Reliable System

By reliability is meant the probability of a device or system performing adequately for the period of time intended under the operating conditions encountered. Thus, reliability implies that the required conditions of operation be known and that a great deal of information about the system be evaluated so as to ensure that it can perform.

Actually, the reliability of the system has been a factor which has been assumed to exist throughout all the preceding steps necessary to obtain a satisfactory system. In fact, reliability must have been included as a factor in establishing the value, determining the cost, and estimating the time, as well as formulating and structuring the system and organizing the method to achieve it. The reliability function is a significant factor in establishing the overall value of the system.[37] Thus, the value of the system is predicated on the reliability having a high enough value. Hence it is necessary for the systems engineer to do, directly and through others, whatever is required to bring about the level of reliability specified.

With the value of systems growing larger and larger, it is apparent that in order to justify the heavy expense of systems of the sizes that are currently being designed and built, as well as projected, such systems must continue to operate a high percentage of the time. That is to say, a system which does not operate satisfactorily may incur losses of significant amounts and tie up large amounts of money which might otherwise be gainfully employed. It is extremely important from cost and safety standpoints for commercial systems to be reliable, and it is also essential for military or space systems to be reliable to provide the measure of safety for the national defense or for the individuals involved. Thus, without question, reliability is an essential consideration both in judging the system and in performing the job of building it properly.

Reliability, of course, is not a characteristic which is easily obtained once and for all. Reliability considerations must be included in a system during its conception as well as during its design phases. Likewise, reliability is determined in no small measure by the way any system of a specific type is designed, by the methods with which a particular system is built, controlled, and tested, by the way this system is packaged and shipped, by the way it is installed and checked out, and by the way it is operated and maintained. Thus, a reliable system must be conceived, understood, and controlled throughout its overall period of existence from conception to operation and maintenance, and maintenance in turn must be performed in a systematic and careful fashion.

By the use of adequate means of recording the field data on prior and existing systems, it is possible to obtain information as to what performance has been obtained from earlier systems and thus to provide the facts to ensure that the current and proposed designs will be adequate and reliable. Furthermore, the matter of reliability is not limited to systems engineering; it is necessary to ensure cleanliness and proper process control, and to diagnose the causes of various symptoms of failure and to initiate and execute corrective action. Furthermore, from a management point of view, it is essential that reliability be considered a prime objective of design and execution, similar to performance, cost, time, and maintainability. Maintainability itself is an important way of ensuring reliability.

Thus, the reliability aspect of systems engineering is a significant factor in the way the overall system job is performed and in the results which may be obtained from the system as operated.

1.7 Conclusions

The system environmental factors described above have served to point out a number of steps which are necessary to bring about the acceptance of a system and its realization. These practical aspects of the overall organization and method of handling a system are quite apart from the analytical tools described in the companion volume, *Systems Engineering Tools*.

The partitioning of the systems environment into value, cost, time, etc., employed has been useful, but it is not the only or necessarily the best order for describing systems engineering methods in the following material. For the majority of the people who will work on or with a system the magnitudes of value, cost, and time will have

Conclusions

been decided before their entry into the system. Most of us are given the magnitudes of cost and time as relatively fixed constraints within which we must operate. We would like to know how the system should be or was organized, how it should be structured, how it can be made to do its job most efficiently. Armed with this general knowledge, we will be in a better position to understand and develop the value of the system, to appreciate and control its cost and time, and to strengthen and enhance its reliability. For this reason, the next chapter will concern itself with the organization of a system so that we may have a better insight of many of the problems of systems engineering methods in general as well as those associated with any particular system.

Although it may seem that these techniques and methods are unique and peculiar to a particular organization or company, they do have common counterparts elsewhere. The fact that an organization may be motivated by its own particular set of values in no way detracts from the fact that all organizations, as well as all individuals, have finite capabilities and finite resources and are limited in various ways which may differ from one to another. These limited capabilities of any organization place constraints on what it can accomplish. Furthermore, people must have ways of evaluating and judging the relative importance of alternative methods of accomplishing things. Therefore, these techniques and methods are of vital concern in the various situations in which systems may operate.

Even if the criteria for a system be as broad as the greatest good for the greatest number, this can be the starting point for systems techniques and methods to take over and provide a framework within which judgment may be made as to which of various systems, perhaps employing alternative methods, will prove most effective for the purpose of the particular organization for which the system is being engineered.

2

System Organization, Scheduling, and Record-Keeping

2.0 Introduction

In addition to technical problems, systems also have organizational and logistical problems. Many different people may be involved over a wide physical or geographic coverage and over a long period of time. Many may work for different companies or organizations with different rules and methods of operating. Very many data and much knowledge are involved. The organizational problem concerns itself with the question of how all these people can work together most effectively for the common purpose. Uniqueness or one-of-a-kind characteristics of a system means that there is probably *no common background;* in fact there are probably many divergent backgrounds so that common words or concepts do not always mean the same thing to different people. Therefore it is important that a simple way of presenting the system to the different people involved be developed, so that they will know what is going on, what is required of them, where they should look for information, who they should provide with information, and in what form the information should be given.

Although there is no one best way of presenting the system (in fact, good ways for different systems may be different), there are common problems for which decisions will be required. It is essential that a realization of these problems be achieved, and that solutions appropriate to the circumstances be reached and explained adequately.

In a number of different areas, it is evident at the outset that such activities as identifying equipment, providing specifications, instituting changes, and checking out of equipment, subsystems, and systems will be required even though their particular form may not be very well defined or understood. An early effort should be made to study the needs of each individual system in terms of its similarities and differences with other systems to determine methods of handling these problems in keeping with the overall system objectives.

2.1 System Engineering Plan of Organization

A systems engineering plan provides for a systematic identification of total requirements on an overall basis. It provides a means for developing hardware, facilities, personnel, and procedural support information on a concurrent and integrated basis, minimizing oversights in design, optimizing design, reliability, and minimizing costs.[36]

Implementation of this plan for a large-scale system requires a major effort utilizing top systems design engineers. Problems imposed by utilizing these key people are small in comparison to the benefits gained. In addition to improving the effectiveness of the overall system, increasing reliability, decreasing downtime, and maintaining cost effectiveness, a great savings is achieved because redundant and wasted effort is avoided. The possibility for procurement of vast quantities of hardware which is ultimately determined to be unnecessary, of stockpiling warehouses with unusable spares, and generating requirements for extensive modification programs can be reduced through implementation of this plan. Great savings which are less apparent will also be achieved, such as more effective personnel training, procurement of realistic quantities of spare parts, proper identification and design of procedural support data, and other support area improvements.

A method of recording the systems engineering planning is used to provide a means of checking, cross checking, and hence quality-controlling development recommendations and the subsequent deliverable items after design recommendations are approved.

A particular overall system engineering plan which has proved effective involves fourteen basic features. This representative engineering plan of organization is patterned after the methods used in the Minuteman Systems Engineering Plan[36] and provides an orderly framework in which the work of formulating and structuring the system can take place. In Section 1.4 the iterative nature of the process of identifying the system objectives and the characteristics of the *operating system* itself were emphasized. In the plan of organization outlined below, the need for also identifying the characteristics of the *support*

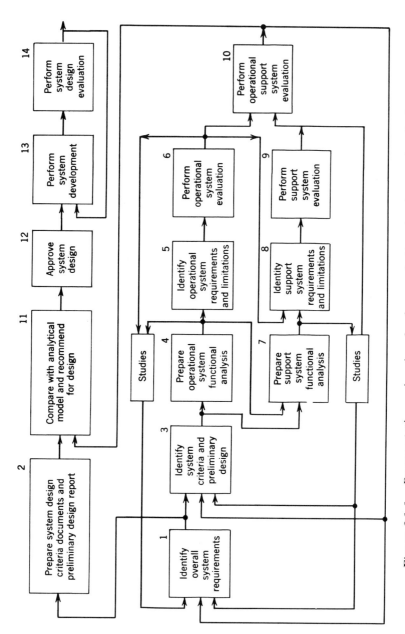

Figure 2.1-1. Representative engineering plan for organization of systems work to be done.

system, reconciling the characteristics of the operating and support systems, and comparing these to the original systems objectives is brought out more forcibly.

The particular plan shown is meant to be a typical form of organization, not a mandatory one. The term organization is used to refer to the items of work to be done and their time order rather than to delineate a fixed administrative grouping of individuals. However, some ideas on the latter topic are included at the end of this section.

Figure 2.1-1 shows in block diagram form the major features of this plan. Inherent in each feature is the need for intelligent thinking, that is, the application of sound engineering design logic, utilizing all known scientific and engineering principles. The fourteen features are briefly described as follows.

1. *Identify the Overall System Requirements, Concepts, and Criteria.* Sources used in this identification will include all pertinent development plans, operational plans, maintenance plans, logistical plans, performance specifications, and design criteria documents. In general, most systems engineering projects that approach the stage where any significant amounts of manpower and money are to be spent on them have had preliminary studies previously performed, appropriation authorizations approved, and overall system requirements set forth. Furthermore, the broad environment of people, resources, and existing conditions will probably be intuitively understood even though not explicitly stated. It is necessary to distill from all this material a concise but complete and consistent set of requirements, concepts, and criteria on which the remaining work may be based. These are generally broad enough so that the changes that may be required from them can be incorporated without modifying the general intent.

2. *Prepare and Record the System Concepts and Design Criteria Documents.* The second step is to determine and record the more detailed criteria and design requirements for the various individual major systems described in 1 above and to publish them as systems concepts and criteria control documents. These concepts and criteria form the basis for a preliminary design report which will later serve as the reference for comparison with the actual system that is ultimately conceived, recommended, and adopted.

3. *Identify System Criteria and Preliminary Design Requirements.* Criteria and design requirements data from 1 and 2 are grouped in such a manner that they can later be translated into functional re-

quirement statements. The emphasis in this step is to subdivide the overall system into major individual systems (or subsystems) and to identify the concepts, criteria, and requirements for each. This is a transitional stage in which the system is partitioned into parts which are more clearly identified with the operational or functional portion of the resultant system.

4. *Prepare Operational System Functional Analysis.* Functional flow drawings are prepared to show in detail the functions that must be accomplished to meet the criteria and design requirements established in 3 above. Pure machine and man-machine interface requirements are identified on the functional flow drawings.

The level of detail to be covered in the functional flow diagrams will necessarily vary from some which are very fine or precise in their requirements to others which are quite coarse or gross, in order to maintain continuity of functions to be accomplished in parallel and/or series. For example, between two gross functions a minor communication function may be required.

No attempt should be made on the flow diagram to mechanize a function in terms of specific hardware, facilities, and people. However, functional requirements for power, signals, cooling, heat, light, sensing, actuation, or other specific conditions should be entered where they are significant.

5. *Identify Operational Requirements and Limitations for Hardware, Facilities, People, and Procedural Support Data.* A system analysis of each functional block on the flow diagrams described in 4 is performed to determine all the technical requirements that are necessary to satisfy the function. The results of these analyses are recorded in reports of a consistent form for use by others working on the system.

Studies of the technical requirements thus established will be made to determine the most reliable, effective means of satisfying these requirements. Decisions regarding whether complete automation is feasible or whether a human interface is required will often involve trade-off studies which will consider system performance, reliability, and cost effectiveness.

All of the recommended requirements are summarized and then analyzed for their impact on facility requirements. If personnel interface requirements are indicated, an analysis will be made to determine whether a need exists for procedural support data, such as operating instructions, inspection instructions, removal or installation instructions, calibration instructions, or theory of operation to permit troubleshooting. Recommendations for meeting these needs will be described.

6. *Perform Operational System Evaluation.* A system functional schematic block diagram showing all the significant end items tied together in a system-installed configuration will be prepared. This block diagram will indicate the flow of signals and/or sequencing requirements and will indicate interface requirements with other systems. This diagram will be reviewed against system criteria and application design principles, standards, and specifications.

Operational time-line analyses and/or loading (extent of utilization) charts will be prepared to show how functions are being met timewise with the use of people and equipment. Critical design reviews will be accomplished to evaluate the system design against system criteria.

7. *Prepare Support System Functional Analysis.* An overall system maintenance concept top drawing will be prepared to show all maintenance conditions to be covered by various individual drawings. Based upon the criteria data of 1, 2, and 3 above, support system functional flow diagrams will be prepared. These flow drawings will cover all scheduled and unscheduled support functions, as well as those support functions which are common to all other related maintenance, to the level of correcting a malfunction of recommended hardware or to performing scheduled maintenance on recommended hardware. In a modeling sense, this may be considered to be a coarse approach to the support system description.

8. *Identify Support System Technical Requirements and Limitations for Hardware, Facilities, People, and Procedural Support Data.* A detailed functional analysis will be completed on the blocks of the maintenance flow diagrams in a manner similar to that of the operational flows in 4 above. The technical requirements imposed by each maintenance function will be entered to the same level of detail as described for the operational systems analysis in 5 above.

Analysis and maintenance engineering will be accomplished to determine the most feasible, reliable, and economic means of meeting the technical requirements as determined by the functional analysis. Trade-off studies will be performed as required, and an intensive effort will be made to minimize system downtime and manpower requirements. Where it is determined that additional technical requirements are needed on the operational equipment to satisfy maintenance—test features, test points, indicators—these features will be added.

The recommended hardware and operational equipment, starting with the systems-installed condition, will be analyzed to determine all technical requirements for maintenance.

A review of the recommended support system requirements will probably lead to such feedbacks as the following:

Redesign of the equipment being analyzed in order to eliminate excessive maintenance requirements.

Addition of maintainability features, such as self-test capability or packaging for ease of replacement, to the equipment being analyzed.

9. *Perform Support System Evaluation.* The system functional schematic block diagram used in **6** above will be revised to show any maintenance engineering additions to the operational equipment imposed as a result of **8** above.

Time-line analyses will be made to show the maintenance engineering required to minimize downtime and manpower requirements. Loading charts will be prepared to indicate the total manpower required, including the extent of utilization and the system downtimes for each maintenance job operation.

10. *Perform Operational/Support System Evaluation.* Calculations and studies will be made from the data of steps 1 through 9 to establish (a) overall system reliability, (b) overall system in-operation rate, and (c) overall system cost effectiveness. These studies and calculations will then be reviewed to evaluate the combined operational/support system design.

11. *Recommend Design.* Upon completing the evaluation in **10** and comparing these results with the desired design criteria of **2**, the technical requirements established may be used to prepare preliminary model specifications, which will recommend the hardware design.

12. *Approve System Design.* The customer review team should study the system recommended for design and approve or disapprove it. Actually, this final review should be the culmination of periodic design reviews with the customer. These should take place at regular intervals which may coincide with significant stages of time or engineering accomplishment.

The output of approved system design summarized above and described in Figure 2.1-1 includes a definition of the individual equipment designs of all the parts of the system and their methods of testing, check-out, and support. Figure 2.1-2 prepared by R. W. Mayer describes the systems engineering process in such a way as to extend the work to include systems equipment development and systems evaluation. Since the results of these two work efforts are

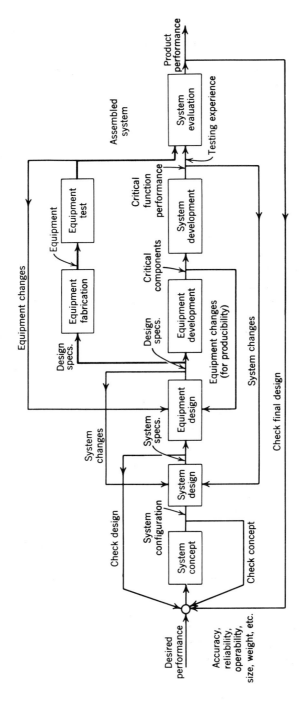

Figure 2.1-2. System engineering process, including equipment and system development and evaluation.

fed back into the system concept, system design, and equipment design, it is important that systems-oriented people be involved in the performance of these added tasks. A description of the work involved in these efforts follows.

13. *Perform Systems Development Leading to Final Specifications.* With the modern methods of analysis available and with the utilization of up-to-date computing techniques and equipment, a major portion of system design can be determined without the necessity of experimental work. However, there are usually certain critical systems functions, the feasibility of which must be checked experimentally by "breadboarding," in a representative physical environment if possible. This work is carried out by systems engineering personnel who use the results to feedback corrections to the system design and to the resulting specifications.

In carrying out this systems-feasibility breadboarding, it should be noted that, wherever possible, components making up systems functions being tested are normally supplied by the equipment design engineers in order that the breadboard may serve to check not only system feasibility but also the adequacy of the specific equipment designs.

14. *Perform system Design Evaluation.* This function is extremely important to the accomplishment of the total systems engineering job in that it closes the loop on the entire engineering and manufacturing process used to produce a systems design for delivery to the customer. Figure 2.1-2 clearly shows this relationship and indicates that it is the function of systems evaluation to ensure that the system design does, in fact, meet the performance requirements of the customer.

This process of system design evaluation is accomplished by such operations as the following:

1. Checking the operation of subsystems when combined in a complete system.
2. Evaluating the validity of assumptions which were made in systems analysis.
3. Paying particular attention to such effects as these:
    a. Cross talk between subsystems.
    b. Power supply coupling.
    c. Mechanical feedback.
    d. Wrong polarities.
    e. Resolution of adjustments.
    f. Serviceability—is it possible to get in and make adjustments?

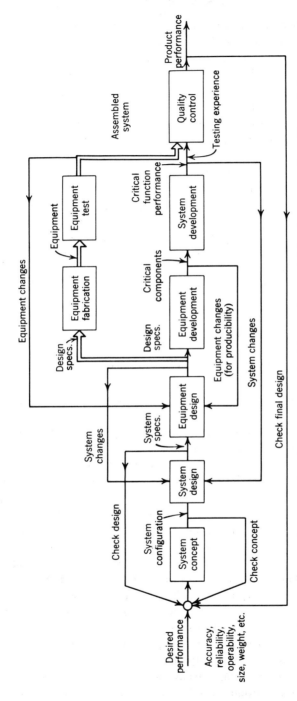

Figure 2.1-3. System engineering process, emphasizing equipment fabrication, test, and quality control.

The above factors are only representative of those which are included in the system design evaluation. It will be recognized by those experienced in this phase of engineering that many of the above factors cannot wholly be evaluated in the initiation of the system design process or may be overlooked in the rush of meeting schedules. By ensuring that systems engineers do, in fact, accomplish this process of design evaluation, it is possible to instill in them a sense of responsibility for these important factors even in the conceptual stage of the system design process.

In complex programs such as are now involved in supplying military equipment, there is normally not time or money to build a complete prototype for design evaluation before delivering equipment to the customer. Instead, evaluation will take place on the first few systems to ensure adequacy of the design. From this point, then, a gradual transition is made from a systems design evaluation to a more quality-control type of testing, which then ensures that the manufacturing process is producing equipment in accordance with the established design. This transition is shown in Figure 2.1-3, where equipment fabrication and test, and systems evaluation by quality control, become dominant factors in the loop.

Systems Engineering Management Procedures

The preceding systems engineering plan of organization has tended to approach the problem of functional requirements somewhat from the viewpoint of the organization performing the systems engineering task. Because of the importance of the systems engineering approach to the user of military systems, the Air Force Systems Command has in recent years[59] set forth its own Systems Engineering Management Procedures, AFSCM 375-5[43] which serve to provide a framework within which to operate on a more common approach to each system. In this way the Air Force is able to benefit from a more uniform set of engineering management procedures from system to system.

The 375-5 procedures are based on a system life cycle which covers a broader time and functional set of needs than those described previously. The procedures are separated into four distinct phases: the conceptual phase (phase 0), the definition phase (phase 1), the acquisition phase (phase 2), and the operational phase (phase 3). Figure 2.1-4[58] shows an overview of this system life cycle.

From this one can observe that phase 0 has as its objectives concept determination and feasibility analysis, as well as the completion of a preliminary technical development plan. The phase 1 objectives include the definition of firm performance requirements that must

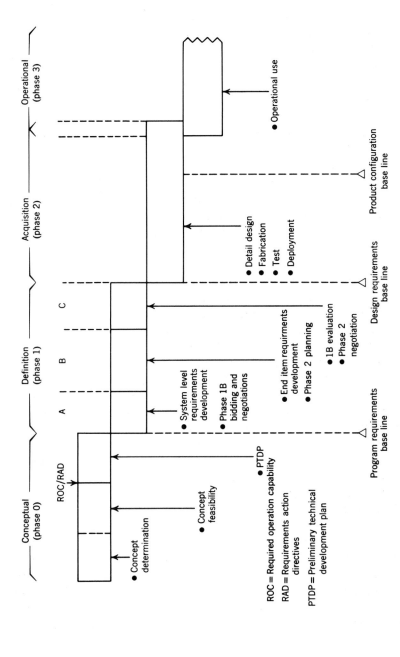

Figure 2.1-4. Overview of system life cycle from AFCSM 375-5.

be met by the system elements, the early identification of significant technical risk areas, and the development of firm schedules and cost estimates associated with the acquisition of the total system. The phase 2 objectives are the obtaining of the system in accordance with previously established and approved requirements; the completion and validation of the definition of the requirements and criteria which are contingent on design; and the development, fabrication, and testing of the system, which is made up of equipment, facilities, personnel and procedures. Phase 3 is the actual accomplishment of the mission objectives by the user with the system that was defined and acquired in the previous phase.

The following material draws heavily from selected portions of the *Air Force Systems Command Manual* 375-5,[43] in which the meaning of the overall system is described and the format for the management of the system is shown briefly in block diagram fashion.

"A system must be designed and tested as a complete entity. The word 'system' has come, through actual practice, to include: the prime mission equipment; its supporting command, control, training, checkout, test, and maintenance equipment; the facilities required to operate and maintain the system; the selection and training of personnel specialists; software for computer program systems; the operational and maintenance procedures; instrumentation and data reduction for test and evaluation; special activation and acceptance programs and logistics support programs for spares and depot maintenance.

"All parts of a system must have a common unified purpose: to contribute to the production of a single set of optimum outputs from given inputs with respect to time, cost, and performance measures of effectiveness. The absolute necessity for coherence requires an organization of creative technology which leads to the successful design of a complex military system. This organized creative technology is called Systems Engineering.

"It must be clearly recognized that in order to satisfy overall system objectives, System Engineering must deal with total system design. As such, it is a unique activity and is not to be confused with the detail design of end items and components. However, it is absolutely necessary that System Engineering management recognizes the predominant and highly complementary role played by engineering design specialists in satisfying the need for total system design. The interplay between the system engineer and the engineering design specialist requires the closest coordination and is a major management problem which must be faced and solved by engineering management.

"System Engineering is fundamentally concerned with deriving a coherent system design to achieve stated objectives. The System Engineering process logically considers and evaluates each of the innumerable military, technical, and economic variables identified by system engineers. Choices of methods of system operation and the system elements are highly involved processes; for a change in one system variable will affect many other system variables, rarely in a linear fashion. The generation of a balanced system design requires that each major design decision be based upon the proper consideration of system variables such as facilities, personnel and training requirements, equipment, procedural data requirements, testing, and logistics within the parameters of time, cost, and performance. This necessitates the closest coordination of select personnel skilled in System Engineering who are to work as a homogeneous system design team. The System Engineering team has the responsibility of translating military operational or advanced developmental requirements into a feasible, economical system. This team responsibility will not be satisfied until system documentation and tests have proven the adequacy of detail specifications for production and performance/design; and have proven that the equipment, facilities, personnel, training, and procedural data provided will satisfy the required military mission.

"No two systems are ever alike in their developmental requirements. However, there is a common and identifiable process for arriving at logical system decisions regardless of system purpose, size, or complexity. This process can therefore be described and specified which relates directly to the management of military System Engineering."

A summary flow of the documentation, major review, and contract action points during the definition, acquisition, and operational phases is presented as Figure 2.1-5 to illustrate the methodology for controlling system engineering effort through the use of the management processes as discussed in AFSCM 375-5.[43] Figure 2.1-5 is divided into three bands as follows: (1) the documentation band, (2) the review band, and (3) the contract band. The documentation band portrays the kind of paper against which review actions will be conducted. The review band illustrates the type of approval activity necessary for implementing a contract action. The contract band shows the contract action resulting from the review activity. A detailed description of each block follows.

Block 1. Preliminary Design Reviews, PDR's, are conducted against the documentation establishing the design requirements base-

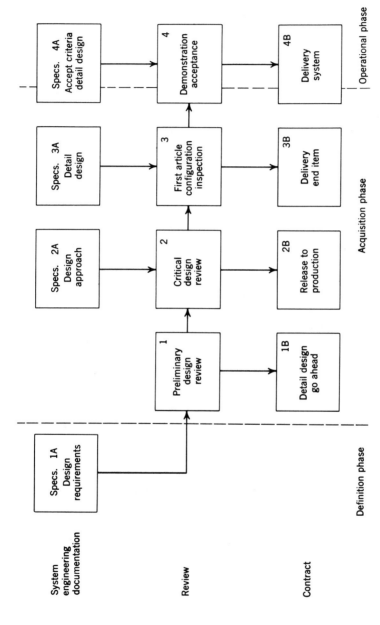

Figure 2.1-5. Documentation, review, and contract action points of AFSCM-375-5 for controlling the systems engineering effort.

line (Block 1A), which consists of design forms and the detail specifications, to ensure that (a) design requirements are valid, and (b) the design approach is acceptable. The result of a PDR is a contract action which permits the contractor to conduct detail design (Block 1B) in accordance with the approved design approach.

Block 2. Critical Design Reviews, CDR's, are conducted against the detail design information available (Block 2A) including the detail specifications, design approach information appearing on design forms, and available drawings. The result of a CDR is a contract action (Block 2B) which releases engineering to production, that is, the contractor is permitted to produce items of equipment in accordance with the detail design presented at the CDR's.

Block 3. First Article Configuration Inspections, FACI's, are conducted against the drawings and the detail specifications (Block 3A) to ensure that the hardware matches the detailed specifications and drawings used in its production. The result of a FACI is a contract action (Block 3B) which accepts the delivery of the first article of hardware.

Block 4. The final action is a demonstration acceptance activity which accepts the assembled system (Block 4B) against accumulated specifications and drawings.

The AFSCM 375-5 systems engineering management procedures are the result of extensive experience in managing the development and use of large-scale military systems. Although these methods should not be considered as final or fixed, they are having a fairly significant effect on the way many large systems engineering organization think and operate.[58] They represent one of the more influential methods for organizing present-day overall systems engineering activity.

Systems Engineering Organization

In setting up most engineering organizations, there are two basic forms that can be considered: the project-oriented organization or the technology-oriented organization. The first subdivides the organization in accordance with the programs or projects under way and applies the necessary skills in each of the programs. The second subdivides in accordance with the technologies or skills required by the various programs. The program type has the basic disadvantage that it makes inefficient use of highly skilled personnel; whereas the technology-oriented type has the disadvantage that it is difficult to pinpoint clearly the responsibility for a program or system.

The systems engineering operation shown in Figure 2.1-6 has four

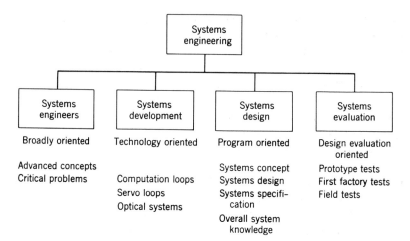

Figure 2.1-6. System engineering organization with flexible functional capability.

major emphases and is a compromise, as are many engineering organizations.

Program Orientation. Program orientation is obtained in the system design unit, in which personnel are grouped in accordance with the systems in the process of design. The engineering leaders of these groups are generally cognizant of all aspects of their particular system in addition to carrying out detailed systems design.

Technology Orientation. Technology orientation is obtained in the system development unit, where personnel are grouped in accordance with skills involved in solving critical technical problems occurring in the design and development of complex electromechanical systems. It is in this unit that analytical work involving the use of digital computation aids or analog simulators is carried out. Some people with broad skills in using electronic, optical, and other devices are located here. Also, feasibility breadboarding of critical functions of the system is accomplished here. Thus, the services of these highly skilled personnel are available for the several different system programs.

Systems Evaluation. The loop is closed on the engineering design process by the systems evaluation unit. This function of evaluating the system design is carried out by prototype testing or by testing the first few units of production runs if time has not permitted a prototype. Also, these people are responsible for technical direction of the field tests required for final evaluation of the equipment.

Systems Engineers.[47, 61] In addition, all three units look to a group of engineers reporting to the manager of systems engineering for expert guidance in the broad aspects of the system, for solution of particularly difficult problems, and for leadership in advanced concepts of system design.

It may be argued that the functions of these various groups of engineers could be more effectively used by organizing on a direct "project" basis with engineers carrying their respective programs through from concept to final test. This may be true provided that most of the engineers are highly experienced in all aspects of systems engineering. However, this is rarely the case, and a few highly experienced leaders must serve as focal points not only for the program but also for the skills involved. Furthermore, as projects start or come to a close, it is possible to use the four functional groups as a basis for a more permanent set of organizational entities.

The organization described above provides this diversity and has been found to work well on complex programs involving guidance and fire-control systems.

2.2 Scheduling and Review

From the organizational and job requirements prepared in accord with the systems engineering approach to the job outlined in the previous section, it is possible to formulate a schedule in terms of specific accomplishments to be realized at certain times. Preparing such a schedule is at best an estimate and very commonly an "experienced" guess. Generally, this undertaking may be approached from two points of view.

The first method is to ask the persons assigned the jobs of doing each portion of the system how long will it take them to do their tasks. A detailed subschedule is prepared for all the jobs making up the required tasks, and an estimate of the total time needed is prepared and submitted to the project or system coordinator. Figure 2.2-1 shows how such a schedule might look as a function of time and serves to point out that some tasks should be started at or before the completion of earlier ones. Also implied is the fact that information to complete a task may be required from a second task that depends on the first task for data.

The second method is to look at the project schedule in terms of its end-date requirements. This fixes the time when it is necessary to have the job done, which in turn establishes when each of the

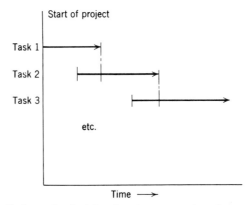

Figure 2.2-1. Estimated schedule of tasks versus time from start of project, from function and equipment viewpoints.

preceding parts of the job must be completed in order to meet the overall objective of the customer and the project. In order to be completed and in operating condition by the prescribed end of the project, the separate equipments need to have been installed, checked out individually, checked out collectively, and operated under controlled conditions for a sufficient period. Figure 2.2-2 illustrates how this approach places emphasis on having time at the end to finish the job and also indicates the need for performing two or more tasks in parallel.

Almost from the first it is apparent that there will be differences in the schedule requirements derived from these two approaches. The equipment designer's estimates are generally longer than can be tolerated by the project requirements, and potential trouble spots or

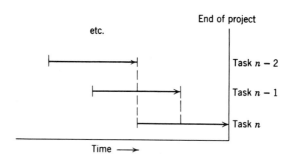

Figure 2.2-2. Estimated schedule of tasks versus time from end of project, from function and equipment viewpoints.

Scheduling and Review

critical problems are frequently revealed from the outset. Changes are generally required in each set of plans, and estimates and revisions in the basic plans and methods may be needed. Figure 2.2-3 shows how the revised schedule has reconciled the times for starting and ending the project and resulted in a set of task times that appear to be reasonably compatible with the designer's and project's needs. Simple PERT charts and more detailed schedules can be prepared from these data as described in Chapter 6.

An important factor in connection with the scheduling of tasks is the identification of specific events, achievements, or goals which and be recognized and agreed upon. Completion of the analysis of a portion of the job, of the list of design specifications for a device, or of the drawings for an equipment is representative of such a specific event. As a means for establishing subgoals and intermediate points in the various subschedules, such intermediate events can be most beneficial.

Two methods of using the schedules and specific events are illustrated in Figure 2.2-4. The first method is to review the accomplishments by sampling each task at equal time intervals to establish whether the job is on schedule or ahead or behind schedule. For jobs of relatively long duration this method is reasonably effective, and PERT or critical path methods tend to use this equal-interval-of-time approach. The second method is to perform the review at a time when the specific event has been accomplished. This, on Figure 2.2-4, corresponds to events a, b, c, and d, only b of which occurs at a time review period. The reviews on an accomplishment basis can be of a more technical nature since at these times a complete unit of work is first available and a comprehensive evaluation of

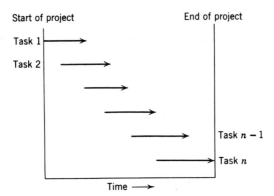

Figure 2.2-3. Reconciled schedule from estimates of Figures 2.2-1 and 2.2-2.

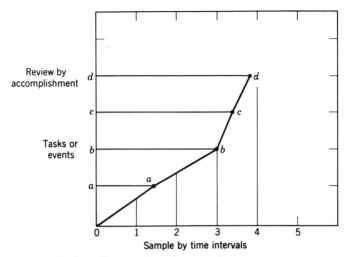

Figure 2.2-4. Review of project based on schedule and task accomplishment.

the whole specific event, accomplishment, or goal can be made more intelligently. Although these subgoals are not always technically compete, in that other events or accomplishments may be needed to provide a "closed loop," each specific event represents the completion of a significant work element and should be given a somewhat independent, even though brief, review.

2.3 Subsystem, Equipment, Assembly, and Component Identification

One of the important requirements of a system is to establish an index or coding system which permits a person to identify readily each subsystem, equipment, assembly, and component as to its location, function, and other features, such as its time phase, from its reference code alone. This indexing or identification scheme should be simple and capable of easy explanation to the many people who will be concerned with the system at the various stages of design, fabrication, testing, installing, operating, and servicing the equipment. Thought and organization are required of the record-keeping and indexing scheme, since in all probability it will be used by human beings as well as by data-processing machines.

There are a number of classification schemes which have been used effectively for large systems, and it will serve the present purpose

Equipment Coding by Location

Equipments in a common system but located at different physical positions can be assigned different major code numbers to identify their location as shown on Figure 2.3-1. Thus, equipments at location A will have numbers 10–19, at location B will have numbers 20–29, etc., for all locations. Furthermore, if it is desired, similar equipments at different locations can have the same second digit in the consistent code of 10's for identifying the location. Thus 10 and 20 might be junction boxes situated at locations A and B, respectively. Similarly 35 and 45 might each represent analog/digital converters at locations C and D. If an alpha-numeric identification scheme is selected, use of a letter or a group of letters may be satisfactory; that is, A, B, C, etc., can serve as the basis for locating the equipment.

In addition to gross location of equipment, identification codes can also be applied to detailed location on the basis of standardized chassis construction practice. For example, J. G. Nish[29] describes how "the design and construction of elaborate electronic systems for instrumentation and control of research and test facilities is guided by principles

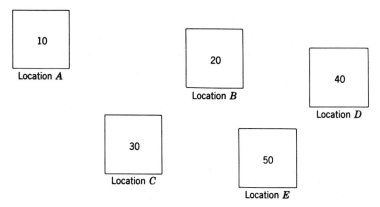

Figure 2.3-1. Identification coding of equipment by numbers and/or letters.

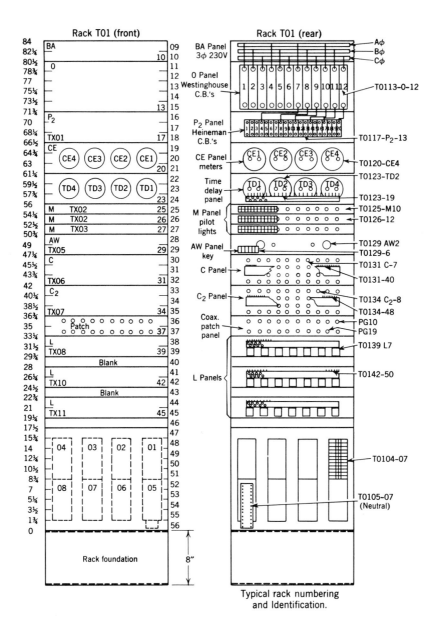

Standardized chassis construction. Modular-height chassis front panels, and numbering scheme to indicate panel location on the rock.

Figure 2.3-2. Standardized chassis construction and numbering scheme.

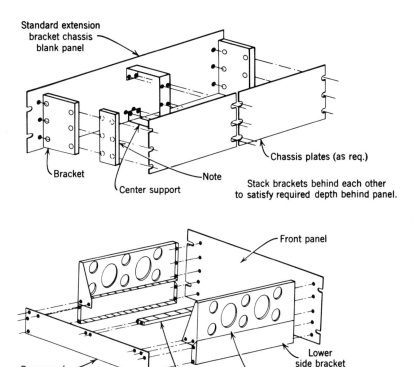

"Erector set" construction of standardized Lawrence Radiation Laboratory chassis from standard, modular brackets and subchassis plates.

Figure 2.3-2 (*Continued*).

and practices of standardization which reduce system cost and completion time." Figure 2.3-2 shows one form of standardized chassis construction, modular-height chassis front panels, and a numbering scheme to indicate panel location on the rack. These panels are all of the same standard 19-in. width but of different heights, each height being made up of an integral number of $1\frac{3}{4}$-in.-high modular units. The rack front panel space is divided into these modular height units, which are numbered from the top down, thus providing a numbering scheme for designating the location of a panel on the rack frame.

To round out and complete the standardization picture used by Nish,[29] a standardized numbering procedure has been adopted so that the number of the rack, panel, or wiring terminal indicates where

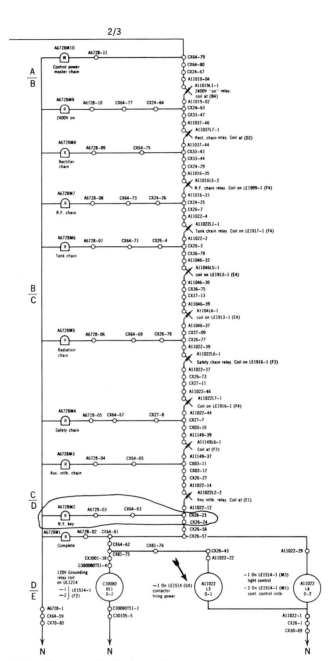

Figure 2.3-3. Lawrence Radiation Laboratory diagram with standardized numbering scheme which indicates at a glance where the rack, panel, or terminal is to be found. Explanation of numbering scheme—Number A 6728-11 (terminal in top line of diagram): (1) first letter (A) indicates area or multiple-rack group in which rack is located; (2) first two digits (67) indicate rack number; (3) next group of two digits (28) indicates panel which occupies modular space 28 on the rack; (4) number (11) after dash indicates eleventh terminal on terminal strip of this panel (panel 28). (Terminal is counted from left end, facing rear of rack.)

Subsystem, Equipment, Assembly, and Component Identification 49

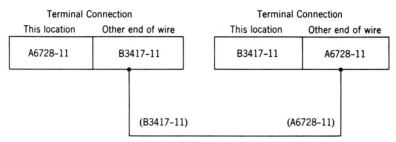

Figure 2.3-4. Identification of location of wire source and destination.

the corresponding rack, panel, or terminal is located. Consequently, the installer or serviceman can tell, from its number on the wiring diagram, where a specific electronic unit or particular circuit connections can be found (Figure 2.3-3). This method has proved to be a tremendous timesaver in installation or servicing.

In another related numbering method which is very effective each terminal location is designated by two numbers, one corresponding to its appropriate geographical location from equipment down to rack, panel, and terminal, and the second corresponding to the appropriate geographical location from equipment down to rack, panel, and terminal of the other end of the attached wire; see Figure 2.3-4. In this way, it is possible for the serviceman to know where to go to check the circuit continuity of the particular wire in question.

System Tree

Another effective way of establishing an index method which facilitates the use of it in interface monitoring, servicing, equipment inventory, and system integration is to code the drawings and documentation by a hardware system block on a common "system tree."

Figure 2.3-5 shows such a proposed system tree for the Apollo system, in which all of the Apollo effort is broken up into a hardware system and a software system. Each hardware system is then divided into its subsystems, and then each subsystem into its subsystems, etc., until when carried to the limit individual nuts, bolts, and pieces of hardware are reached. The depth of level; for example, 1, 2, 3, etc., of the "tree" and the codings system A, B, C etc., are dependent on the detail required in carrying out the various tasks.

Another example of the use of the "system tree" and the structure coding concept is shown in Figure 2.3-6, which is taken from a space vehicle "failure analysis report."

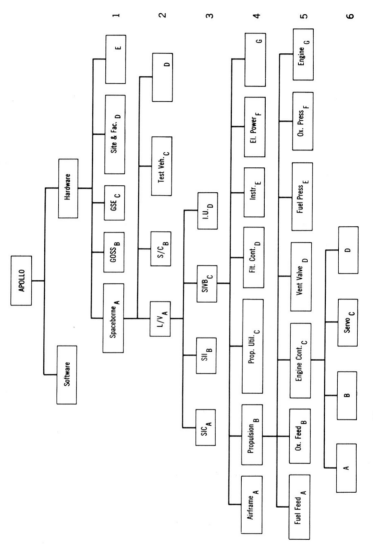

Figure 2.3-5. Tree method for indexing system parts.

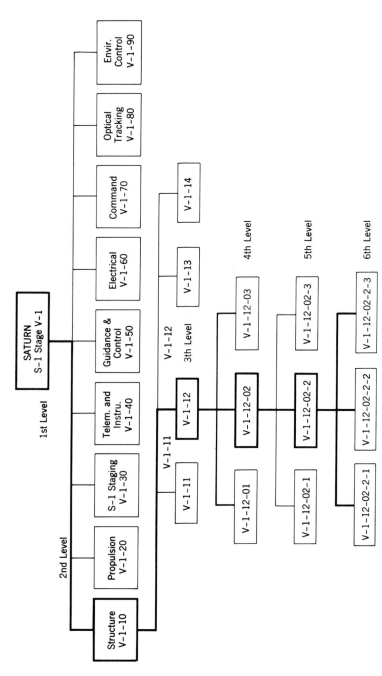

Figure 2.3-6. Typical structure code used by MSFC for identifying systems as referred to in "failure effect analysis" report.

As seen from the previous two examples, a system tree which is properly constructed can help greatly in the following goals:

1. Defining and organizing identifiable functional elements of the system.
2. Defining the needed drawings, specifications, and documentation for the system.
3. Establishing common equipment terminology.

2.4 Record-Keeping and Communications[34]

A major factor that makes systems engineering different from individual problem solving is that many communications problems exist. The nature of these problems is apparent if the systems engineering process is visualized as an information network in which one portion represents the systems engineering group and the other parts represent groups such as management, design engineering, manufacturing, the customer, research, and others that act as sources or sinks of information with respect to the systems engineering group. In this network, two-way communication links should exist between each pair of groups involved.

If one could observe the flow of information in each of the links from the beginning to the end of a number of projects, he would see a definite pattern in the kind and amount of information flow and records required. These observations might suggest the question: Given the network, what can be done to the patterns of flow in the various links to maximize the efficiency of the system? In particular, what kinds of records are needed by the various groups originating these records, and what can be done so that the groups which can use this information have ready access to it? For example, the information gained during the development and design of an equipment determines the performance and the variation in performance with various design parameters and can be very useful later on during the process of manufacture as well as during the test and check-out phases.

The record-keeping scheme is best organized in terms of its overall needs, and a consistent set of all records, including drawings, specifications, equipment description, and test data, should be compiled and readily available. In addition, circuit diagrams, system word descriptions, drawing sequence lists, assembly drawings, interface one-line diagrams, and interface listings and status should be included in the system of records.

Information regarding test performance at different stages of manufacture and assembly should likewise be available in a consistent set of identification numbers so that it is possible for people using the equipment during check-out or normal operation to have access to these data acquired previously on the equipment. Furthermore the inclusion of a failure report system in the process of record-keeping is highly desirable so that difficulties encountered in the field can be quickly fed back and the records become available to the designers in terms of the experience actually encountered with their designs. This will, of course, make more complete the reliability records on each of the items of equipment.

It is of considerable importance that the methods of identification of components, assemblies, equipments, subsystems, and systems described in Section 2.3 be employed to permit easy storage and retrieval of information over widely separated geographical and organizational groups. Standard design procedures, recommended people to see on various kinds of problems, recommended practices, and references to specifications and procedures should also be included in the record-keeping procedure.

Table 2.4-1 indicates the type of record-keeping that has proven valuable in some large-scale systems. It is intended to be indicative, rather than definitive, and does not show in any case the magnitude or form of the material involved. This is particularly true of the measuring-instrument calibration curves; many records have curves which are generated as a result of the calibration associated with testing. The necessity of transmittal of this type of curve to the operating areas is obvious.

The necessity for feedback of failures and design discrepancies to design groups would appear to be obvious, but it often happens that some organizations may provide a sufficient buffer between test and design groups to obscure details of difficulties. The need for correlating failures with past and future occurrences provides an impetus for keeping records of this sort; care must be taken to ensure that complete information is currently available to satisfy the correlation required. A design discrepancy should, of course, be remedied immediately, and subsequent testing should clearly establish that the design is now adequate.

Two other areas of feedback are suggested which, although they are important, are easy to overlook. The "as-is" schematic drawings go with the equipment and represent to the operating group what hardware is actually delivered to them. This, in itself, helps to reduce the possibility of error introduced through other documentation chan-

Table 2.4-1. *Record-Keeping Activities*

Information Source	Type of Information	Type of Record	Feedback to/from		
			Design	Test	Operation
Component failure	Data/method of failure	Failure report	x	x	x
Design discrepancy	Corrective action required	Unsatisfactory condition report	x	x	
Operational system calibration	Data	Test report	x		x
Measuring instrument calibration	Data	Calibration curves		x	x
"As-Is" drawings	Current status	Schematics Drawings, sketches		x	x
Daily activities	"Ghost" troubles, sequence of test difficulties	Daily log test report		x	x
Peculiar troubles	Any	Report	x		x

nels. The test report on daily activities provides a historical source for certain questionable areas which appear during system operation. If a history of a persistent "ghost" trouble exists for a certain system, then time must be taken to exercise it. Of course, the whole problem should have been solved in the preshipment testing. However, because of the persistent conflict between check-out and shipment dates, if one is inflexible, the other will suffer when unusual conditions arise.

Nothing can replace personal communication between test and operating people and the use of the telephone as a reporting tool. This person-to-person contact does not eliminate other methods of reporting, but rather supplements them and rounds them out considerably.

2.5 Check-out and Servicing

A lot of time, money, and manpower training is involved in the check-out and servicing of equipment. It is therefore worthwhile

Check-out and Servicing

to spend the time and effort necessary to see that these procedures are standardized to the maximum extent possible to permit simplification in methods and equipments as well as the time of testing. This standardization, must, of course, be compatible with the performance obtained and the cost required. A number of studies have been made to determine the feasibility of standardizing test equipment effectively.[28] The purpose of these programs was as follows:

1. To eliminate unused capability and the overspecification of test equipment accuracies.
2. To simplify complex testing operations.
3. To reduce the size and weight requirements of organizational maintenance.
4. To incorporate principles of human engineering.

A detailed investigation of the characteristics in terms of the input-output signal parameters of a large number of electronic test equipments provided the means for direct across-the-board comparisons and showed that signals to be generated and monitored by test equipments were common to most systems. To provide the basis for the comparison of similarities and the detection of important differences between test equipments, a method was evolved which describes the major components and subassemblies in terms of their input and output characteristics and power requirements. Units have been specified through technical characteristics by either of two methods: identification by transfer functions, or delineation of input and output signals.[28]

For the first analysis input-output characteristics were grouped by wave-shape types as follows:

1. D-c, constant.
2. D-c, variable.
3. Sine waveform.
4. Peaked waveform.
5. Rectangular waveform.
6. Trapezoidal waveform.
7. Complex waveform.

Determination of the density of signal wave types indicated that sine waves dominated, constituting 37 per cent and 32 per cent of all imputs and outputs, respectively. In second place were d-c signals with 27 per cent and 25 per cent; rectangular, peaked, and complex waveforms each accounted for about 10-15 per cent.

As an indication of the fact that check-out can be done on a fairly organized basis, Figure 2.5-1 shows how the number of cases of check-out requiring voltages in the various ranges indicated can be grouped. Because of the nature of the power supplies which provide energy to the electronic equipment, it is apparent that not all ranges of voltages are required and that the eight voltages indicated plus the five unknown voltages are able to cover the input-voltage requirements for a large number of cases. The approach used in handling the check-out problem should be to assure that testing methods recur and that they can be identified and categorized according to their purpose. After this is done, it is necessary to derive data from the actual tests performed in the various categories. In this way each category or test type can be reduced to a minimum number of test configurations having the maximum number of capabilities. Figure 2.5-2, showing stimuli input waveforms, indicates seven different forms of voltage stimulus with the functions and waveforms indicated.

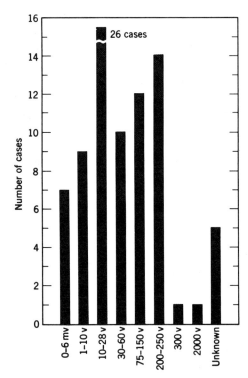

Figure 2.5-1. D-c input voltage requirements—number of cases versus voltage ranges.

Stimulus	Function	Waveform
1. D-c voltage	E_{dc}	
2. Sine waveform	$E \sin \omega t$	
3. Sawtooth waveform	$E(\sin x - \frac{1}{2} \sin 2x + \frac{1}{3} \sin 3x - \frac{1}{4} \sin 4x \ldots)$	
4. Square waveform	$E(\cos x - \frac{1}{3} \cos 3x + \frac{1}{5} \cos 5x \ldots)$	
5. Pulse waveform	$E[k + \frac{2}{\pi}(\sin \pi k \cos x + \frac{1}{2} k \pi \cos 2x + \frac{1}{3} \sin 3k\pi \cos 3x + \ldots)]$ where $k = \frac{\text{pulse width}}{\text{period}} < 0.1$	
6. Amplitude-modulated waveform	$E[1 + m_a F(t)]$ where $F(t)$ is any specified stimulus function	Where $F(t)$ is sawtooth
7. Frequency modulated waveform	$A \sin [\omega t + m_f F(t)]$ where $F(t)$ is any specified stimulus function	

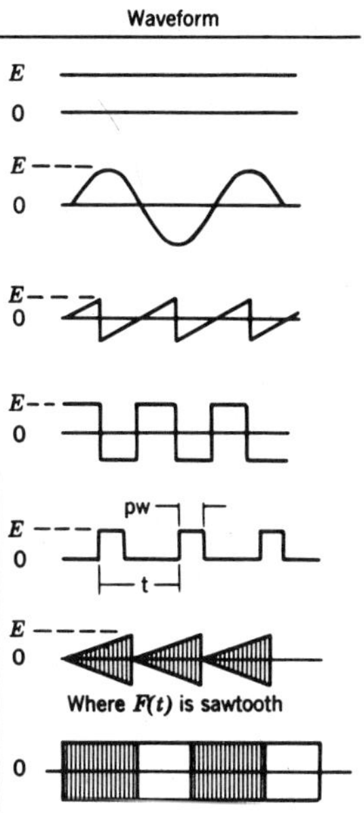

Figure 2.5-2. Stimuli input waveforms.

A tabulation of the occurrence of test types reveals that the tests were almost equally divided into two major groups. Some of the more significant figures from the test type investigations were as follows:

Test Types	Total Test Analyzed (%)
Monitor	
Resistance	10
Voltage	20
Monitor-stimulus correlation	
Gain	10
Frequency response	10.5
Wave shape analysis	8.0

The similarity between these five test types suggests that similar methods might be applied to all. The last four types are so similar that they can be performed with the same kind of test equipment configuration, using a variation of the frequency response technique to simplify the operator's part in the test.

Automatic Testing

Automatic testers have been used extensively and have demonstrated that they can significantly reduce the time required for testing. That automation can be incorporated into a test system has been shown by test sets built to demonstrate that functional modular standardization of circuits and modules is practicable.

Manual functions, together with test systems components to perform these automatically, are shown in Figure 2.5-3. An automatic test system consists of five basic functions:

1. Programing equipment to provide controls.
2. Switching unit to route signals.
3. Test evaluator to determine the test results.
4. Adapters to link the test system and the unit under test.
5. Stimuli to provide signals for the unit under test.

Selection of one type of function is determined by accuracy, environmental conditions, characteristics of the unit under test, and the purpose of the test—whether for operability or fault isolation. Equally important and interdependent with the selection of function is the selection of the test method. Static testing (point-to-point tests made with signal stimulus) permits continuity, leakage, impedance, transformer ratios, and point-to-point voltage tests, but does not measure frequency response, power outputs, or noise, or detect oscillator drift, intermittent operation, etc.

Dynamic testing is performed on energized equipment with stimuli applied. This may be open loop (a fixed signal is applied and a measurement is made of the reaction of the unit under test) or closed-loop (the input stimulus is dynamically varied as a result of the feedback).

Logistics

It is evident that the problems of supply, maintenance, and training are all interrelated with test equipment capabilities. The need for various quantities and types of equipment is dependent on the fact that the same test equipment can be used for each of these different purposes. In a review of a number of systems which might be expected to be supplied by a common set of test equipment, it was found that a total of 53 different types of test equipment modules would be required. Figure 2.5-4 shows a cross plot of the quantity

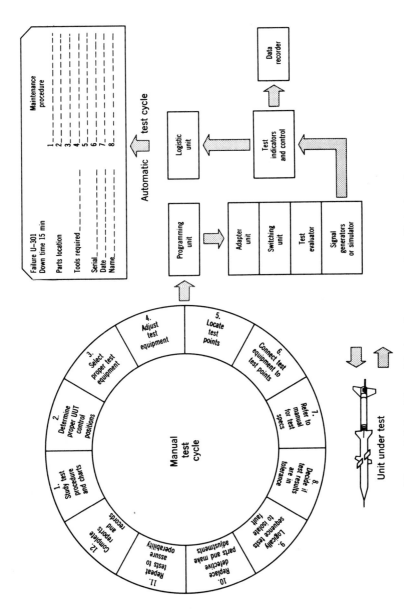

Figure 2.5-3. Manual test cycle versus automatic test cycle.

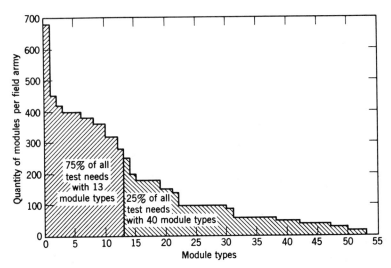

Figure 2.5-4. Distribution of standardized test equipment modules.

of modules versus the module types. It indicates that by using a small number of modular types, 13 in this case, 75 per cent of all the test needs could be met, whereas 25 per cent of the test needs require an additional 40 modular types. Whereas these data are applicable to the check-out and test of a particular set of systems, such a functional approach to the problem of test-out, check-out equipment will probably indicate for other systems the benefit to be gained through standardization.

The results of an intensive study of check-out equipment have established conclusively that the modular concept of assembly in the form of standard building blocks can be developed for test equipment. This modular concept was not limited to general-purpose test equipment items but applied also to special-purpose equipment, for which the advantages of assembly from standard on-the-shelf modules were proven. The basic premise of sub-equipment-level standardization was validated. Test assemblies developed from the standard modules simplified operation, reduced control and indicators, were self-calibrating and monitoring, and utilized automatic testing techniques.

Through standardized check-out and servicing equipments and methods, it has been found that:

1. Efficient specialized test assemblies can be constructed from standard electronic test equipment developed in accordance with the criteria and techniques determined during study phases.

2. Test assemblies possess minimum complexities and maximum expansion capabilities.
3. Modular test assemblies can achieve a weight and size reduction of from 30 to 50 per cent over existing equipment.
4. Modular test assemblies can be cheaper, more reliable, and simpler.
5. The lag time in test equipment delivery can be reduced by approximately 50 per cent.
6. The number of test items to support existing systems is considerably less, of the order of 50–60 per cent, than would otherwise be the case.
7. Modular test assemblies are compatible with manual, semiautomatic, or fully automatic operation.
8. The operational simplicity of testers developed permit more effective maintenance with fewer personnel of lower skill and less training.
9. Modularization permits immediate introduction of transistorized circuits.

The check-out and servicing of complex, electronic systems are of extreme importance from a time, cost, and personnel point of view. It is essential that an organized procedure for handling this problem be considered early in the program and that methods be undertaken to resolve questions concerning manual, semiautomatic, and automatic testing, the level and frequency of testing, and other such items to achieve the greatest effectiveness for this portion of the system requirements.

2.6 Similarities and Differences

In keeping with the concept that systems change with time, that is, are evolutionary, it is important to consider each system, in particular the current one, as it relates to previous and probable future systems. In this regard it is significant to determine in what ways this system is similar to previous ones and in what ways is it different. In considering these similarities and differences, it is worthwhile not only to keep in mind the technical objectives or the basic methods employed, but also to consider all phases of the operation both from a human resources point of view and from a technical equipment point of view, including manufacturing, test, and other facilities. For example, it may well be that the main similarity between the current system and the previous one is the fact that the same group of design engineers has worked on both systems. Another similarity may be that the equipment will be built in the same area and the manufacturing facilities, test equipment, personnel, etc., are geared to the particular way of handling things that was used in the past. Another point of similarity may be in the nature of the people using the equipment. Here, training and facilities which have been invested in the past may be most effectively used with only relatively minor changes in the circuitry equipment and other materials being employed.

On the other hand, in the matter of differences, the introduction of new techniques, such as transistor circuitry versus tube circuitry, may require a completely different method of handling the instruction and the matter of servicing than has been previously used, even though the same nominal circuit functions—amplification, modulation, demodulation, etc.—are employed. Here, the use of transistor voltage levels and circuit concepts will invalidate most of the concepts and equipments previously employed. It is important in the planning and organizational stages that these differences be noted and proper attention be paid to accommodating for them. Certainly to the extent that it is possible to take advantage of existing instruction books and testing courses, as well as actual training equipment, efforts should be made to include this type of logistics in the organization and planning of the operation.

It suffices to say here that in the organization of a new system a particular effort should be made to determine points of similarities and differences with those of previous systems to the extent that they can be employed to advantage. Capitalizing on the similarities may well result in decreasing costs and reducing the possibility of major errors. To the extent that differences exist, these differences may represent problem areas requiring considerable effort to ensure that these potential sources of trouble, high cost, and time-consuming delays are not allowed to become dominant obstacles. For these reasons, similarities and differences represent a focal point at which particular attention is required.

2.7 Specifications[32]

Specifications in engineering serve much the same function between two sets of parties as do contractual or legal agreements between two sets of businessmen and/or lawyers. They provide a set of statements which indicate what the systems engineer requires of the various equipments and subsystems in terms of their various characteristics and features so as to permit the overall system to obtain its required objectives. In turn the specifications serve as a set of references against which the designer, manufacturer, and tester can check their work with the mutual understanding that if each subsystem and equipment can meet its specification the overall system will meet its specifications and objectives.

Therefore, specifications serve as a powerful systems tool to decouple the work of the systems engineers and the design engineers, production group, and quality-control and test engineers. Without such specifications, a "chicken-and-egg" situation exists. The systems engineer cannot work out his design because the subsystem and equipment characteristics are not known, and the subsystem and equipment characteristics cannot be designed because the system requirements or specifications are not known. With such specifications, however, each group can proceed about its own work, with a reasonably clear understanding as to what the other is expecting to do and what overall results will be obtained when the subsystems and equipment have been produced to these specifications.

In order for this desirable situation to exist in fact, the system engineers must have studied and analyzed the requirements of the system, subsystems, and equipment to be able to draw up the specifications correctly. Also, the remaining design, manufacturing, quality-control, test, and other groups must be sufficiently competent and skilled to perform their assigned tasks with confidence of being able to meet the specifications to which they have agreed.

Needless to say, the mere act of drawing up specifications and agreeing to them is not sufficient to guarantee that either of these difficult-to-obtain conditions is met. Furthermore, as with business contracts and legal agreements, specifications in general are a source of never-ending discussions, arguments, and reinterpretation. However the presence of specifications does permit the system needs to be stated and the capabilities of the subsystems and equipments to be expressed in a common language. Hence, specifications provide an information interface which is extremely valuable for successful system operation.

Early Specifications

Since for many new, one-of-a-kind systems, there exists a mandate to go ahead with the system before its exact characteristics or detailed objectives are known, the availability of a set of specifications or the information with which to generate such a set cannot be immediately realized. Rather than wait until a complete set of correct specifications can be issued, it is important to start the system design with as many data as are available and to indicate what form the specifications will have without necessarily specifying the final quantitative values.

In order that the design may proceed in an uninterrupted fashion, the systems specification must be a continuous "living document." This can be accomplished by issuing, very shortly after the start of the program, specifications which show in only a general fashion the concept in its embryo form. These first specifications may be no more than an outline with blank spaces, but they start information flowing from system design to equipment design and thus permit the build-up of people and information relative to the equipment characteristics and requirements. Also, in the same manner, initial development plans may show only broad areas of test and analytical work.

The early issuance of such preliminary plans and system specifications serves to acquaint the equipment design engineers with the problems that will be arising in the future. In accordance with these early specifications they may initiate preliminary equipment designs. This approach also provides the equipment engineers with a chance to feed back information into the systems area at an early date, thus closing the feedback loop. In this process, the systems concept can be revised at appropriate intervals in accordance with the difficulties of the equipment engineers in meeting their requirements.

Firm-up Specifications

In order to finalize the specifications from which equipment will be built, both analytical work and feasibility tests must progress in the system and equipment areas. Only when a sufficient amount of this work has been accomplished and enough information has been generated can the systems specifications be sufficiently complete to make final the equipment design and fabrication. Thus the loop containing equipment fabrication and test (Figure 2.1-3) is allowed to proceed. In order for equipment designs to proceed at the earliest possible time, systems specifications should indicate, as they evolve, the firmness of certain portions of the design. With this information the equipment engineers can go ahead with the design of the firm portions of the system even though the overall design may be incomplete.

Changes to Specifications

After the system specification has become sufficiently firm to warrant the start of detailed equipment design, provision must be made

Design Changes 65

for formally changing the specifications in a minimum of time. A system of issuing specification change memos has been found effective to provide a means of documenting these system changes quickly and in an organized fashion.

The specification change memo is used by the system engineer to document an agreement between himself and the appropriate equipment design engineer. The change involved may be a system design change, which may require an alteration in equipment design, or an equipment design change, which may necessitate an alteration in system design. The latter may occur in the design stage when it is found that excessive equipment cost or complexity may be incurred in meeting the original system specification. Thus, the feedback loop containing the design processes involving iterative changes in the system design and equipment design is made a tight one, available, as required, by the use of these procedures.

Specification Format

Although there is no universally accepted format for engineering specifications, a general sequence of topics that has been used quite extensively and satisfactorily has grown up around military specifications. This sequence stresses other generally applicable specifications as to materials, procedures, and practices; specific performance requirements; and specific test procedures and methods. Environmental conditions, painting, packaging, shipping, and other requirements are also specified. A list of typical specification headings is given in Table 2.7-1.

2.8 Design Changes

In keeping with one of the major system concepts, it is inevitable that there will be changes in the design of an equipment or subsystem. These changes may be relatively minor to accommodate small improvements in methods of manufacture, materials employed, or test procedures. Instead, however, they may be fairly major, for example, a complete change-over from one equipment style to another. Obviously, the nature of methods of incorporating these two types of changes may differ, and different procedures for handling such changes may be required. In the case of minor design changes, there will probably be no question about any changes in specifications, but for a major modification, specification changes may be required. A parallelism might be drawn here between the specifications for design

Table 2.7-1. MIL-S-6644A (USAF) MILITARY SPECIFICATION: *Specifications, Equipment, Contractor-Prepared, Instructions for the Preparation of* . . .

1. SCOPE
    1.1 SCOPE
    This specification covers the instructions to be followed for content and form of specifications to be prepared and furnished by a contractor when specified in Air Force Contracts for items of equipment for which no suitable specification exists.
2. APPLICABLE DOCUMENTS
    2.1 The following documents form a part of this specification:
    SPECIFICATIONS
    STANDARDS
    PUBLICATIONS
3. REQUIREMENTS
    3.1 General Form and Content
    3.2 Section 1—SCOPE
    3.3 Section 2—APPLICABLE DOCUMENTS
    3.4 Section 3—REQUIREMENTS
    This section will cover a major portion of the specification and includes subsections on

    | Preproduction sample | Radio-interference suppression |
    | Components | Dimensions |
    | Material | Weight |
    | Design | Finish |
    | Construction | Government-furnished property |
    | Performance | Government-loaned property |
    | Details of components | Workmanship |
    | Interchangeability | |

    3.5 Section 4—QUALITY ASSURANCE PROVISIONS
    This section includes subsections on
    Quality control through surveillance

    | Classification of tests | Acceptance tests |
    | Test conditions | Test methods |
    | Preproduction testing | Parts destroyed in test |

    3.6 Section 5—PREPARATION FOR DELIVERY
    This section includes such subsections as

    | Preservation and packaging | Physical protection |
    | Intermediate packaging | Marking |
    | Packing | |

    3.7 Section 6—NOTES
4. QUALITY ASSURANCE PROVISIONS
5. PREPARATION FOR DELIVERY
6. NOTES

and the constitution of an organization, in that both are changed much less frequently than are alteration notices and the corresponding by-laws of an organization, which are meant to be more flexible.

Alteration Notices

Alteration notices are a convenience set up between equipment design engineering, equipment fabrication, equipment test, and quality control to permit minor changes in the design, manufacture, or test of an equipment in a fashion that is known and mutually acceptable to all parties concerned. Generally, these notices are drawn up rather informally and are reviewed at periodic intervals, say weekly, by an alteration review board which then authorizes or rejects the incorporation of the changes into the equipment design, fabrication, and test procedure. Since such minor changes are inevitable, the method of building systems should be designed to be flexible enough to incorporate them without requiring major decisions by those responsible for the overall conception and operation of the system.

Major Modification Changes

Major modification changes may be required to introduce into the system improvements of a significant nature. These changes may be essential to overcome certain deficiencies in the equipment or subsystem design that have been revealed, or they may merely be desirable because they incorporate cheaper, lighter, or easier-to-manufacture equipment.

These changes, obviously of a more extensive magnitude, should be introduced only after careful attention and planning. An effort should be made to schedule or time them in such a fashion that there is a minimum disturbance to the overall system. They may represent a rather severe transient in the system's life cycle and produce conditions analogous to instability or poor control in a conventional position control or servomechanism.

When the Mod change is required to overcome a system deficiency, there is less control over the time when it should be made. A balance must be struck between the danger involved in failing to introduce the change, and the damage which will result if the change is introduced too soon, that is, without sufficient preparation and planning.

When the Mod change is merely desirable because of the improved characteristics of performance, cost, etc., that will result, the time of introduction is more amenable to control, and an effort should be made to group a number of changes together to form one overall

Mod change. An effort should be made to have this Mod change coincide with other major schedule changes or shut-down or maintenance events that are planned. Here again, the gains to be realized from introducing the change sooner should be balanced against the losses incurred if the change is introduced before it is well thought out or debugged.

It is not proposed here to describe the methods for introducing modification changes, for many different ways exist and have been found effective. The purpose is to point out that major modification changes are almost inevitable, and that a good systems method will include provision for satisfactorily handling these changes as a routine procedure. In this way, the time required to plan such a modification change is expended when the pressure and anxiety of an actual change are not present; therefore the planning of methods can be more thorough and will take less time from implementing the change than if initiated only after the necessity arises.

Check List of Problems and Possible Solutions

It is inherent in the systems engineering situation that a number of activities are taking place concurrently with interrelated time and performance objectives. Frequently the situation originally envisioned or planned for a particular activity is different from that actually encountered. Hence, new problems are continually arising for which general, long-term solutions would be desirable but for which there is no time to devote at present. Also, there are minor items to be taken care of when the current short-time objectives have been met.

By jotting down a list of such problems when they become apparent, so that they are available to mull over or solve when the time exists, one is able to take care of a lot more of them than would otherwise be the case. Dr. C. F. ("Boss") Kettering was one who claimed he used this "ten-best-problems list" concept to help him be the prolific problem solver that he was. By having a file or list of suggestions for improvements and changes, or of just plain inconvenient things that are happening to the system, to be used by all persons connected with the system, one is much more likely to find solutions quickly.

Frequently, being able to formulate or state a problem is a major step toward its solution. A problem check list, especially with some ideas for remedying or solving the difficulties, can be of considerable help to the overall improvement and success of a system.

2.9 Conclusions

The process of engineering a system should itself be handled in a systematic fashion. By organizing the plan of activities to be accomplished in engineering the system, one can emphasize both the characteristics of the operational system itself and those of the support system with its design implications which are necessary to install and maintain the operational system. Scheduling from both the system and the equipment points of view is important and can serve as a basis for periodic and functional reviews as the system development progresses. Equipment identification and record-keeping can be a great benefit to improved communication within the organizations that are making and will later use the system.

Check-out and maintenance support equipments may represent a significant investment of time, money, and manpower in operating a system. However, they may be able to produce significant savings if their design is approached in a standardized, modular fashion.

Specifications can provide a basis for better understanding among the various groups working on the system. In order to prevent the slowing down of early design efforts, it is desirable that the initial specification objectives be allowed some flexibility and a method for change if this is needed. Likewise methods for incorporating design and materials changes must be provided so that the system may be able to accommodate the changes that take place in its environments.

3

Formulating and Structuring the System

3.0 Introduction

The processes of formulating and structuring a system are important and creative,[9] since they provide and organize the information which, for each system, "establishes the number of objectives and the balance between them which will be optimized." Furthermore, they help identify and define the system parts which make up its "diverse, specialized structures and subfunctions."[48] Hence, formulating and structuring a system provide methods for relating (1) what the system consists of in the mind of the persons or group desiring it; (2) what it means in terms of the persons or group designing and building it; and (3) what it will provide to the persons of groups operating, using, and servicing it. They provide a set of "reasonable" parts and methods of relating them so that the many persons working on the system can understand the whole in sufficient detail for their purposes, and their particular parts in explicit detail so that they may contribute their best efforts to the extent required. A further purpose of system formulation is to recognize the magnitude of the job, including the possible pitfalls.

Formulating consists of determining the system inputs, outputs, requirements, objectives, and constraints. *Structuring* the system provides one or more methods of organizing the solution, the method of operation, the selection of parts, and the nature of their performance requirements. It is evident that the processes of formulating a system and structuring it are strongly related, and for that reason this Chapter will consider them together.

Seldom is there complete agreement between various individuals or groups as to what constitutes the "correct" formulation or structuring of a system; however, it generally appears that the number of really good system formulations or structures which can be agreed

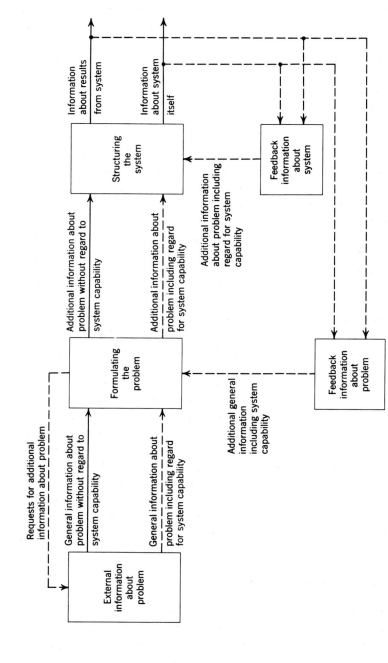

Figure 3.0-1. Information flow diagram showing relationships between formulating and structuring and other aspects of the systems problem.

upon at a particular time is relatively few. The point of significance here is that the process of formulation of the system problem is not generally a unique one and that it pays to do the best one can with the formulation process and move on to the process of structuring. After looking at both the formulation and structuring possibilities, it is then time to reformulate and probably restructure further.[20] This approach is to be preferred over one of providing a more lengthy formulation of the problem followed by an independent structuring effort with little feedback or iteration between the two processes.

Figure 3.0-1 shows how the *external information about the problem to be solved* initially provides some general information about the problem (solid line) without very much regard to the capability of the system that is required to solve the problem. This serves as the input to *formulating the problem.* The output, or result of the formulation process, is to provide additional detailed information about the problem, again in all probability without detailed regard to the system capability. It is this information which serves as a major input to *structuring the problem.* The structuring process has as its output *information about the characteristics of the system* that has been synthesized as well as *information about the results which would be obtained from such a system.* With these data from structuring about the system and its resulting system characteristics, it is now possible to *feed back information about the system and the problem* both in general and in detail to influence the formulation of the problem and the structuring of the system. As the formulation of the problem is proceeding, it is frequently necessary to request additional information about the problem from external sources. The process may be somewhat likened to the game of twenty questions, in which additional information is sought by further questioning to develop and clarify the idea of the answer being sought. An output of this question-and-answer process of formulating and structuring a problem is to bring out what is not known but is required for a satisfactory solution to obtain the system being sought. Knowing the obstacles is frequently essential to being able to overcome them.

3.1 Problem Objectives and Goals

It has often been said that, if a person knows his objectives, he is well on his way toward meeting them. Certainly in control systems the importance of having a strong and accurate reference signal or statement of control logic is very apparent. In formulating system

Problem Objectives and Goals 73

objectives, the need for clearly defining them is equally important, but the problem is complicated by the fact that there are a number of objectives—performance, cost, time, etc. In addition, these objectives exist with different importance and values for various people and groups. Thus, those concerned with defining the system, with formulating and structuring it, with building it, with operating it, or with maintaining it may all have assigned to them different, but related, objectives and goals in each of these categories. Furthermore, since these groups may have different capabilities, skills, and other commitments, the significance of the various objectives may differ appreciably among the groups. It is essential therefore that not only the objectives of the overall system be established and defined, but also the objectives of the various subsystems or functions involved be set forth.

As has been mentioned in Section 1.3 of *Systems Engineering Tools*,[49] one generally acceptable set of objectives and goals includes performance, cost, time, reliability, and maintainability. Whether the goals and objectives are these or some others, it is essential that they be expressed in terms as specific as possible. First of all, what order of magnitude is reasonable and proper for performance? Is it 10 per cent, 1.0 per cent, or 0.1 per cent accuracy that is being sought? Are the rates required 10 per second, 100 per second, or 10,000 per second? Although the person responsible for solving the problems may not know the performance requirements well enough to be willing to specify them, nevertheless, if asked, he is generally quite *willing to indicate what he is **not** thinking of*. Hence, by elimination of unreasonable answers for some of these requirements, one is able to infer quantitative values which can be agreed upon. It should be noted in passing that, when "pressured" into committing himself to a given set of numbers, a person will not always come up with the "right" set. Hence the accuracy of data thus obtained may be considerably in error, and other ways of checking these quantitative requirements should be sought if at all possible.

One factor of considerable importance in establishing objectives is to determine *which* of the various *objectives appear to be* most *important*. Is cost, time, performance, or reliability the main problem? For some space problems, reliability may have an overriding precedence; for some utility or industrial systems, cost or time may have top priority. This relative weighting can be of considerable significance in establishing the systems objectives.[9] Not only are the relative weightings in terms of importance significant, but reasonable values for magnitudes of these quantities are also of great interest.

Someone will say that he needs the equipment early; he means that he needs it before something else which is not needed for one year. Likewise, he may have high performance specifications which are not consistent with the cost he is willing to pay. When confronted with a choice of meeting either specifications or cost, he is faced with the reality of his needing the performance and the price is permitted to rise, or the cost is fixed and he will accept the best performance possible for that price. Trade-off knowledge of this sort can be very valuable to the systems engineer in the process of problem formulation and structuring.

Frequently a set of proposed specifications can be very useful in helping to establish system objectives. However, more often than not such specifications are not available. If this is the case, the systems engineer may find it extremely helpful to *draw up* **tentative** *system specifications* which the customer can review and either accept or change as appears appropriate. Frequently this tentative set of specifications will serve as the basis for a more complete and final set that can be mutually agreed upon by customer and systems engineer.

This specification-writing "exercise" can be of great help in establishing in the mind of the systems engineer what he really thinks is important and what is its relative significance to other things. In addition, it may serve a similarly valuable function for the customer. Here again, knowing what facts cannot be agreed upon may itself prove to be most informative. It is well to keep in mind that preliminary specifications should endeavor merely to cover the more important general requirements and to provide room for later changes if these become necessary. At this early stage it is generally not possible or desirable to spend too much time writing specifications, but it certainly is desirable to have some common basis of agreement between customer and systems engineers. However, as mentioned in Chapter 2, it will be necessary for more complete and detailed specifications to be prepared later when such information is available.

One other aspect of system objectives that is helpful in the problem formulation stage is that of describing the nature of the inputs and outputs to each of the different organizations involved in the system. Figure 3.1-1 shows in matrix form how each organization function, such as synthesis and analysis, will have its own input-output set of requirements and in turn may contribute to other functional organization requirements. Here again not only will information factors be involved but also cost, significant time dates, and other interrelated dependencies may exist and constitute constraints that must be consid-

System Environment

Organization Function	Inputs	Outputs
Synthesis		
Analysis		
Design		
Manufacture		
Quality control and test		
Operation		
Maintenance		

Figure 3.1-1. Matrix showing systems organization functions versus their inputs and outputs.

ered sooner rather than later. Data such as these can serve as a basis for an activities-flow diagram from which later PERT diagrams can be drawn.

Although setting up commonly agreed upon objectives and goals is not an easy task, it is essential in minimizing later misunderstanding and in providing a realistic basis for a common effort by all groups concerned. It is important at the early stages of this process for one to realize that his efforts are tentative and are subject to such changes as reasonably may be required. When the major objectives are established and appear to be capable of attainment, they should serve as quite firm requirements and considerable effort should be made to achieve them.

3.2 System Environment

The environment of a system is in reality a special set of inputs that give the particular system a frame of reference or character which is relatively difficult to change or which changes relatively slowly with respect to the principal inputs being manipulated. The environment may include such factors as the following:

1. *The physical environment:* The size, power, and energy requirements of the system's output objectives; or the temperature, pressure, humidity, vibration, and shock conditions within which it operates.

2. *The state-of-the-equipment (or art) environment:* What asso-

ciated equipment can be changed, what cannot be changed, and why? What are the characteristics of the equipments or systems that supply this system? What are the characteristics of the equipments or systems that this system supplies? What is the state of the particular art—is it accelerating, uniformly changing, or static? What constitutes the limits of what can be accomplished and at what price or time?

3. *The organizational or human environments:* What are the organizational characteristics within which the system must operate? What appear to be the skills and desires of the different people who ultimately are to use the system or its outputs? How much change in organization and people appears to be possible or desirable?

4. *The time and money environment:* What is the time span within which the system must operate or be put into action? How much time does it take to provide the inputs, and how quickly or slowly can the outputs be used? How much money is available for the job?

In contrast to mathematics, in which the form of the solution is of greatest significance, systems engineering is most concerned with the relative size or importance of things. Therefore it is essential to understand the values of magnitude so as to be able to judge their relative importance. Thus in some systems given variables will be considered as inputs, whereas in other systems these same quantities will be considered as environments because of the characteristics or relative interests of the particular system in regard to certain types of phenomena.

Figure 3.2-1 shows in a pictorial fashion how both inputs and en-

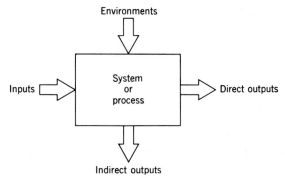

Figure 3.2-1. Inputs, outputs, and environments to system and/or process.

System Environment

vironments enter a system or process. Each will in general contribute to both the direct and indirect outputs. Thus,

$$\text{Direct outputs} = f_d \text{ (inputs, environments)}$$
$$\text{Indirect outputs} = f_i \text{ (inputs, environments)} \quad (3.2\text{-}1)$$

where the functions f_d and f_i will probably not be the same.

Before proceeding with a more detailed discussion of environment, it is important to realize that to change or condition the environments may be easier than to incorporate within the system the ability to cope with various values for the different environmental factors. The system engineer should be willing to consider as one or more of his alternative solutions the possibility of changing the environment through controlling it to some extent.

Physical Environments

One of the principal effects of changes in environment over a period of time is to alter the gain or functional relationship between input and output. Thus, if

Output $= f$ (input, environment), to a first approximation

$$\Delta O = \frac{\partial f \text{ (input, environment)}}{\partial \text{ input}} \Delta I \text{ (input)} \quad (3.2\text{-}2)$$
$$+ \frac{\partial f \text{ (input, environment)}}{\partial \text{ environment}} \Delta E \text{ (environment)}$$

Equation 3.2-2 shows that, in addition to changes in output caused by ΔE, the $\Delta O / \Delta I$ ratio is directly influenced by the absolute value of the environment through the $\dfrac{\partial f \text{ (input, environment)}}{\partial \text{ input}}$ term.

For example, with

$$F = ma \quad (3.2\text{-}3)$$

the acceleration that is realized for a given force is

$$a = \left(\frac{1}{m}\right) F \quad (3.2\text{-}4)$$

If m changes 2/1, the acceleration produced by a given force is ½. See Figure 3.2-2.

Regulation of Physical Environment. Since changes in environment may produce a significant effect on the operation of a system or its

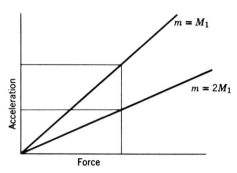

Figure 3.2-2. Acceleration-force relationships for different masses.

parts, it is frequently desirable or necessary to isolate the local environment from its natural state and to regulate it before using it. Figure 3.2-3 shows how an unregulated environment is acted upon by a regulator to produce a more nearly constant environment.

Although only one stage of regulator action is generally required, it is of course possible for there to be two or more stages if necessary. Figure 3.2-4 shows two such stages of environment regulation. The first stage provides coarse regulation, with perhaps 1–5 per cent of the variations of the unregulated environment. The second stage provides fine regulation with perhaps only 1–5 per cent of the variations of the coarsely regulated environment. The overall improvement for the part of the system of interest may be simpler and cheaper with two regulators than with only one regulator that is required to do an overambitious job. Table 3.2-1 shows how energy variations in a number of different forms, such as electrical, mechanical, and thermal, are regulated and the nature of the relative magnitude of the variation of the inputs and the corresponding output regulations. Although reduction in variations of 2/1 is generally easily obtained, and 100/1 is quite common, 250/1 or greater, is usually obtained only at premium price or effort. Frequently the inherent limitation

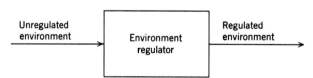

Figure 3.2-3. Effect of environment regulator on unregulated environment to produced regulated environment.

System Environment

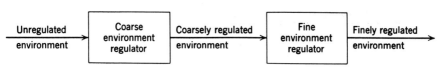

Figure 3.2-4. Use of two stages of environment regulation.

turns out to be the inability of the basic sensor to measure the variable being regulated to the desired accuracy.

The other side of the environmental picture is the conditioning of raw system outputs to bring them into useful form. Figure 3.2-5 illustrates some means that are employed for conditioning outputs of different energy forms.

Other Environment Considerations. Other environment considerations refer to such general limitations as the power or material requirements, as well as the nature of the disturbances, upsets, and noise considerations which serve to provide constraints on the level of performance achievement which can be realized. Although frequently not mentioned initially or specifically, these factors provide a real limitation as to how good performance can be reasonably expected to be.

A prime example of such other environment considerations is the power requirement of the system. The useful power needed to perform the desired system function, for example, to move the load,

Table 3.2-1. *Comparison of Typical Input Variation Regulator Type, and Output Variation for Various Forms of Energy Devices*

Energy Form	Input Variation	Regulator Type	Output Variation
Electrical	Voltage $\pm 10\%$	Regulating transformer	$\pm 1\%$
		Voltage regulator	$\pm 0.25\%$
	Frequency $\pm 5\%$	Speed regulator	$\pm 0.1\%$
	Load $\pm 100\%$	Current regulator	$\pm 1\%$
	R-f noise ± 1–2 volts	Electrical filter	± 5–10 mv
Mechanical	Vibration $\pm \frac{1}{2}$ in.	Vibration absorbers	± 0.001 in.
	Shock ± 10 g	Shock mounts	± 0.5 g
Thermal heating	Temp. change		
	$+2000°C$	Heat shield	$+50°C$
	$\pm 50°C$	Heat exchanger	$\pm 1°C$
Hydraulic or pneumatic	Pressure ± 500 lb	Accumulator	± 20 lb
	Flow $\pm 100\%$	Regulator	$\pm 1\%$

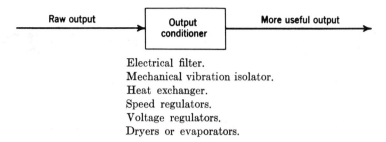

Electrical filter.
Mechanical vibration isolator.
Heat exchanger.
Speed regulators.
Voltage regulators.
Dryers or evaporators.

Figure 3.2-5. Means for conditioning outputs for different energy forms.

may be only one third to one tenth the power rating of the motor required to move itself as well as the load. The inertia of the load may be a small part of the inertia of the motor, and the power required of the motor must supply the combined motor and load inertia. This problem can be approached first by determining *the desired output power requirement* and then by determining the *total system input power requirement* necessary to produce the desired output power.

Thermal and mechanical environments frequently have different characteristics over different conditions of operation—short-time ratings, long-time values, shelf-life conditions; all these may provide certain significant environmental factors that help to shape the overriding environment for the system.

The magnitude and frequency of occurrence of disturbances, such as mechanical torque or vibration disturbances caused by adjacent equipment, may provide important environmental inputs. As has been previously noted, "every system is really a subsystem of some larger system," so that occurrences in other parts of the larger system must be considered as they may provide a significant environment for the particular portion of current interest.

State-of-the-Equipment or State-of-the-Art Environment

The equipment environment refers to the associated equipment with which the equipment presently being considered must operate compatibly. The state-of-the-art environment refers to the degree of development of the particular field of activity or equipment at the present time or with presently available techniques.[48] Perhaps, each of these ideas can be thought of as a conscious recognition of the fact that other portions of the "larger system" of which this system is a part change slowly or only with great expense. A major decision

System Environment

is whether, for the system being considered, one wishes or is able to afford to make a major expenditure of time, money, or effort to achieve the goals that are apparently desired. To make this sort of decision, it is necessary to have relevant data and to understand what "break-points" or conditions of discontinuities exist in this system situation. The extent to which existing equipment should remain fixed and the extent to which it can be modified or replaced should likewise be understood, preferably at the outset but in any case as soon as possible.

In addition to the normal existing-equipment and existing-methods inertia which is generally present to limit change, other fundamental restrictions exist. Figure 3.2-6 shows a generalized cost versus size-of-equipment curve which is divided into three regions, normal, miniature, and oversize. The term cost is not necessarily limited to financial cost but may represent effort, time, or any other value function. In the normal region the costs are proportional to the size, and there is no particular reason for selecting one size or another on a cost/size ratio basis.

Below a certain size, for the environment in which the judgment is being made, a condition may be reached where to make things smaller means that special care, treatment, handling, etc., are required and the costs rise very rapidly. This miniature region is one in which an inverse cost/size relationship appears to exist so that costs increase as size decreases. In a trade-off sense, one is no longer paying for the material in the equipment but for the labor, time, or special effort involved.

There may also exist an oversize region in which the number of

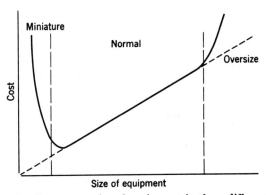

Figure 3.2-6. Cost versus size of equipment in three different ranges.

equipments of the size required is so small or the features involved in making the item are so unique that again special handling or design may be required in addition to the increased material costs which exist. Here again the cost/size incremental rate is higher than normal, and the question should rightfully be asked, "Is this particular oversize requirement necessary?"

From a systems point of view, it is highly desirable to know whether one is in a miniature, normal, or oversize region of operation, i.e., environment, and to what extent one is paying a premium for the particular size of equipment that he thinks he needs.

Figure 3.2-7 illustrates the fact that standardization and/or manufacturing limitations frequently cause the pricing to increase in steps rather than in a direct proportion as shown by the dotted line. After one has established whether the required equipment is being costed on a stepped rather than a proportional basis, he should then determine whether there are reasons for not trying to use equipment at the high end of the step. Thus, once a size greater than I on Figure 3.2-7 is required, it is worthwhile to consider going to size II to provide added capability or reserve without an increase in cost.

Although it can be said that the state of the art is gradually improving with time, we can see from Figure 3.2-8 showing the discovery of chemical elements versus time that the normal improvements (plateau regions) tend to occur at a slower rate than that which prevails during breakthroughs when radical changes take place and large improvements are made in a short period of time. In our present fairly scientific society, it is generally the case that these breakthroughs take an unusual amount of effort to achieve.

Whereas it is by no means certain that one can achieve a sought-for

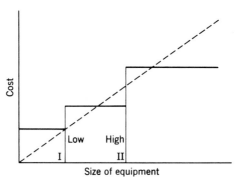

Figure 3.2-7. Cost versus size of equipment for standardized-cost sizes.

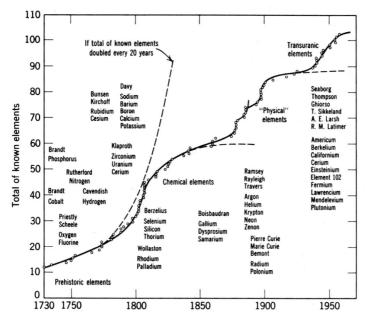

Figure 3.2-8. Chart of chemical element discovery, showing that each period of scientific growth is followed by a period of relative inactivity lasting about three times the growth period.

breakthrough, at any particular time the degree of improvement is to some extent a function of the time which has elapsed since a breakthrough has occurred. Referring to Figure 3.2-9, one notes that in region A, presumably following a breakthrough, additional input effort yields a somewhat proportional increase in useful output. After

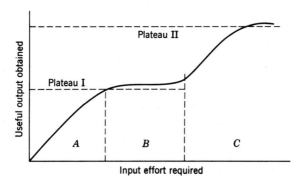

Figure 3.2-9. Useful output obtained versus input effort required.

the more direct applications of the breakthrough improvements have been realized, further increases in useful output require considerably greater inputs, such as are shown in region B. If a new breakthrough is achieved, region C, then again a high rate of change of useful output results.

From the systems point of view, it is desirable to know whether one is in a technical environment where an increase in effort is likely to yield a small increase (B) or a large increase (A) or (C) in output.

Organizational and Human Environments

The degree of acceptance and success realized by a given system is frequently materially affected by the people who have to use, service, maintain, and otherwise work with it. The "human time constant" may be the longest one in any system; changes in people may take place very slowly. Therefore it is highly desirable to know the framework of the organization as well as the equipment within which operate the human beings who will be affected by the proposed system. Project orientation or functional orientation of the individuals involved may make a difference in the ease of system acceptance. Likewise other habit patterns or skills may influence the amount of effort that is needed to implement one system approach versus another.

Another environmental consideration is what the objectives of the group desiring to use the system are (or appear to be). Occasionally the ultimate objective is not the one initially considered as the objective, and sometimes an understanding of how people really will use the system may yield a considerably different formulation of it. A case in point might be an information storage and retrieval system for library use. The answer to the question of whether it should be primarily useful for trained librarians supplying data to technical people or for trained technical people unskilled in library arts who will need the information ultimately may result in two radically different approaches to the way the information system is designed.

The questions of how much of an organization exists, how willing the people who may be involved are to change, how much novelty in organization appears to be required for the possible system configurations are environmental factors worthy of consideration.

The fact that the organizational and human environments should be incorporated into the system formulation by no means implies that they should be kept inviolate. However, due allowance for what they are should be made, and such changes as may be required must be included in the system plans.

Time and Money Environments

The amounts of time and money considered appropriate for an equipment or system are important factors which influence many of the decisions that have to be made in a system. Whereas these magnitudes may be reasonably well founded on fact or careful engineering planning, it may also be the case that they are not compatible to the other factors involved. It is most worthwhile to establish the approximate values for time and money and to determine their relative effects on other system variables.

Figure 3.2-10(a), (b), and (c) show as a function of time a set of curves for the achievement, its rate of change, and the rate of change of achievement rate, respectively. Although the general shape of these curves will differ for each system effort, essentially three well identified areas of interest are readily recognizable: the starting period, when the acceleration is high and positive; the high-rate period, when the achievement is great; and the stopping period, when the deceleration (negative rate of change of achievement rate) is high.

Since the starting period is one of relatively small manpower and poorer definition of the problem, the achievements are difficult to measure and expenditures are hard to identify clearly as to their ultimate worth. Frequently this initial period can be quite decisive in terms of interpreting the system to the customer, and the need for bold new approaches is timely.

In the high-rate period the initial decisions have already been made and the major amount of work is taking place. Expenditures are high, and careful scrutiny of each job is difficult. Minor changes are being made as the need for them becomes clear, but the designs are essentially fixed.

The stopping period occurs when the job appears to be done in the sense that all the parts are essentially finished, although unfortunately the equipment doesn't work yet, a lot of the people who started the job are leaving it without having fully trained others, and a number of newcomers have arrived who know only a part of the problem. Since the expenditures probably are close to their budget value or are well over it, the need for looking at additional costs very carefully exists. Frequently schedule demands and other requirements place great emphasis on getting the work done regardless of cost or of number of men or overtime required.

As in the previous cases, the need to understand the conditions of

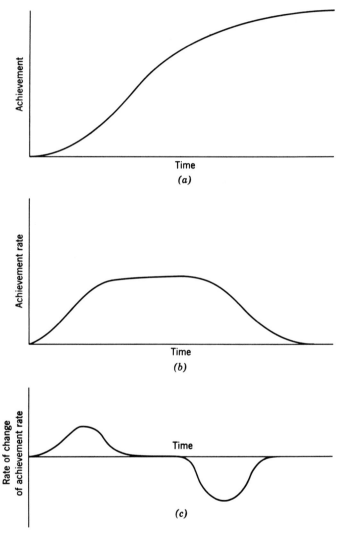

Figure 3.2-10. Achievement, achievement rate, and rate of change of achievement rate versus time for a system project.

the problem is great. Since the time and money environments may play a fairly dominant role in making systems engineering decisions, these environments should be known not only in general but also in a fair amount of detail so that due regard can be taken of the opportunities that may exist.

3.3 What Are the System Inputs?

The inputs to a system include both controlled and environmental variables and can be divided in an approximate fashion into signal and power inputs. The signal inputs are concerned primarily with their information and control characteristics, while the power inputs are involved mainly with their energy or material characteristics. Whereas the form and timing of the signal inputs are important, the energy level and efficiency of the power inputs are most significant.

Some of these system inputs are ones which are dictated to the designer by external systems or conditions over which he has little control. These he must learn about in detail to help him determine how their characteristics will affect his system as well as how his system may affect these inputs. Other system inputs are ones that are selected by the designer and are used as internal signals; these he should select so as to be convenient from such considerations as standardization, cost, or performance. Thus, for the external signals, the designer must learn of these and adapt his system to them. For the internal signals, the designer has the option of selecting them to best serve his purposes.

Figure 3.3-1 shows a generalized system embodying a process and its control.[22] The signal inputs labeled *external signal inputs* and *external signal disturbances* may be ones over which the designer has little or no control. The power inputs of *process power and/or material sources* and *process disturbances* likewise may not be subject to his control. On the other hand the equipments that supply the *internal signal inputs*, the *intermediate signal variables*, the *intermediate process variables*, the *actuating signals*, and the *output signals*, as well as the equipment that performs the *control logic* and the *control instrumentation and conditioning* are ones over which he has a far greater voice in the choice of their characteristics. The selection of the *control power sources* and the *control disturbances* may or may not be his direct responsibility and would be handled in a fashion appropriate to the degree of choice enjoyed by the system designer.

To be able to describe and identify the signal and power inputs, whether from external sources which he must understand from someone else's description, or for internal use which he must specify to subsystem designers, it is important that the system designer have available commonly understood nomenclature and definitions. The following material serves to illustrate first some of the signal source

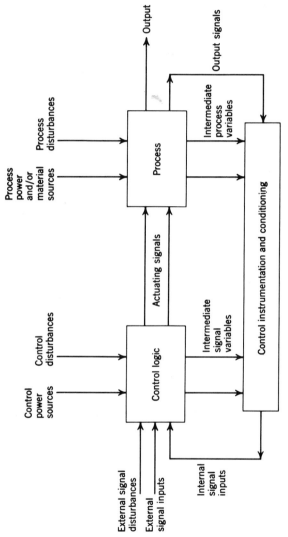

Figure 3.3-1. Generalized system embodying a process and its control.

What Are the System Inputs?

characteristics of importance and then some of the power source characteristics.

Description of Signal Source Identification Terms

Signals may be either analog or digital in nature with a range of hybrid analog-digital signals in between. Existing ASA definitions and symbols should be employed wherever possible to permit the users to have more ready understanding of the terms, based on generally accepted application of the definitions. ASA C 85.1-1963[39] has gained wide support for many of the terms that follow.

For analog signals such identifying terms are

Gain or sensitivity: the equivalent value of the signal quantity received expressed in terms of the corresponding value of the quantity which causes it.
EXAMPLES volts per degree, degrees phase angle per inch, inches per gallon per minute, cycles per second per volt.

Accuracy: the amount by which the indicated value of the signal may differ from the true value of the quantity it represents.
EXAMPLES 0.1 per cent of full scale or 0.5 volt whichever is greater, 2 per cent of indicated reading.

Deadband: the range through which an input can be varied without initiating a response, expressed in per cent of maximum input or in absolute magnitude.

Saturation: the limit of the magnitude ratio between output and input at which any further change in input no longer results in any appreciable change in output.

Linearity: the closeness with which a plot of input versus output approximates a straight line. Linearity is usually expressed as a nonlinearity, that is, a maximum deviation from linearity, or the departure of the actual slope from that of the straight line.

Hysteresis: the property of an element evidenced by the dependence of the value of the output, for a given excursion of the input, upon the history of prior excursions and the direction of the current input.

Maximum and minimum amplitude: magnitude of largest and smallest values capable of being produced.

For digital signals identifying terms include the following:

Codes: the equivalent value of the signal received expressed in terms of the symbol it represents.
EXAMPLES binary, decimal, binary coded decimal, octal, gray, alphanumeric, etc.

Repetition rate or frequencies: the maximum speed or frequency of the medium carrying the information or the carrier on which it is transmitted.

EXAMPLES 100 kc 200 bits per second, 10 increments per period.

Signal equivalency: the condition which represents the one or the zero of the binary code.

EXAMPLES $+6$ volts $= 0$; 0 volt $= 1$; return to zero $= 0$, nonreturn to zero $= 1$.

Other signal source characteristics which are significant and independent of whether analog or digital data are used include such factors as the following.

1. *Dynamic characteristics:* the nature of the range of the magnitude of the input variations with time that must be capable of being accommodated. These variations may be expressed in terms of stepfunction inputs, specific sinusoidal inputs at various frequencies, rates of change and/or higher derivatives of input with time, spectral densities, autocorrelation functions, or other appropriate functions.

2. *Input impedance characteristics:* the internal regulation properties of the input equipment which are affected by the characteristics of the load that the system presents to this equipment. Typical of such descriptions of the impedance are its time constants, its bandwidth, or its output impedances whether these be in electrical, mechanical, hydraulic, or some other form.

3. *Coordinate reference systems:* the nature of the set of axes relative to which are measured the signal quantities that serve as inputs to the system. Examples include pressure measurements relative to ambient, absolute temperature measurements, angular rate measurements relative to body bound axes, line of sight angles relative to inertial space.

The above defining and descriptive terms serve to indicate the nature of the detailed signal source, identifying terms which are necessary to formulate the specific system problem being worked on. Although by no means complete, they also indicate the fine detail of information required to handle the multidimensional problems that tend to occur in systems engineering.

Description of Power Source Identification Terms

On Figure 3.3-1 were shown power sources for both the control and the process. Their role was one of providing the environment for these functions. Hence it is necessary to be able to describe quantitatively the characteristics of these power sources in terms that can readily be interpreted.

Typical of such power source identification terms are the following:

1. *Nature of primary power:* the energy form from which is derived the primary power for the system. Examples might include commercial electrical supplies, diesel engines, steam turbines, gas turbines, chemical batteries, or solar batteries.

2. *Magnitude and power level:* the power rating on an average, peak, and short-time basis. If the power is electrical, the voltage level, the type of current (direct or alternating), and the frequency all must be stated.

3. *Regulation:* the extent to which departures are to be expected from the nominal values for the various quantities stated. Examples might be ±10 per cent voltage magnitude variations, ±5 per cent frequency variations, ±50 per cent speed variations.

4. *Dynamic response:* the nature of the time response of the power source to abrupt changes of load. This might be expressed in terms of magnitude and duration of the power dips as a function of load, that is, a transient response to a step disturbance. It might be expressed as a frequency response or bandwidth capability.

5. *Reliability and life:* terms which can be expressed in many ways. Mean time between failure, availability, shelf life, and hours of use, are typical.

These and many other terms are available for identifying the power source characteristics and are required at times so that a proper formulation of the system problem can be accomplished.

Disturbances and Noises

In addition to the inputs that are purposely entered into the system, a number of unwanted inputs or possibly upsets to the system generally affect its operation or its outputs. Rather than be surprised that these unwanted signals (or noise as they are sometimes referred to because of their similarity to unwanted acoustical noise) exist, the system engineer should make a conscious effort to look for possible disturbances and to identify their magnitude and time characteristics. Generally there is some lower limit or threshold below which it is not possible to go in measurements or sensing. It is important to be aware of any such boundary value and to determine whether it is above or below the level of accuracy or performance desired.

Typical of disturbances and noise that may prove objectionable are such phenomena as the following:

Mechanical friction. This may be static or coulomb friction or perhaps dynamic or viscous friction that is a function of speed.

Unbalance forces or torques. Unsymmetrical weight distribution, firing forces not directed through the center of gravity, aerodynamic lift forces that vary with speed and do not pass through the center of gravity, and product-of-inertia terms that couple between axes are examples of such unbalance forces and torques.

Variations in power sources. Changes in power or signal voltages, speed, pressures, and flow attributable to other equipment on the system or on adjacent systems for which variations may occur that are independent of anything happening to the principal system involved may be considered as being variations in the power sources.

Signal-to-noise ratio. The magnitude of the signal amplitude relative to the comparable noise amplitude is a factor of importance in describing how much extraneous information, i.e., noise, there is relative to the maximum (or average) signal. Since the system frequently cannot distinguish between signal and noise, this ratio helps to determine the lower limit below which the system error cannot be conveniently reduced.

Electrical interference or noise. This can generally be expressed in terms of its amplitude, i.e., rms value, and in addition its time characteristics, i.e., spectral density or autocorrelation function.

The choice of terms—signal sources, power sources, and disturbances and noises—may be somewhat arbitrary, and there may not always be agreement as to whether an input is one or another. The name is not nearly so important as the magnitude, time characteristic, or effect of the input. It is essential that all the purposeful inputs as well as the unintentional or unavoidable ones be expressed. Not only need there be given nominal values for the various quantities, but also a range of tolerances that represent reasonable magnitudes of departure from the nominal values.

3.4 What Are the System Outputs?

The system outputs can be considered to be of two principal types: those which are essential and should have certain desired characteristics, and those which are nonessential or even detrimental and should provide limited amounts of these undesirable characteristics. Although our immediate approach to describing the system outputs tends to emphasize the required ones, it is frequently of comparable importance to be equally specific in delineating the undesirable characteristics which must be minimized.

Also of significance in describing and specifying the required system output is a consideration of what will be done with these outputs, how will they be used, and what characteristics of the person or equipment using the system output may affect the desired output or its characteristics. In keeping with the concept that each system is in reality a subsystem of a larger system, it is desirable to determine the use to which the system outputs will be applied and to consider them as system inputs to the next system. From the input needs of the second, or next, system, what would be the best form for the outputs of the first system?

For example, in information systems the amount of data which is required for display to an operator may be small compared to the detailed data which are necessary to perform later calculations. Whereas the displays can be simple, the other data may be stored semipermanently, for example on tape, to quite an extensive degree and in a form eminently well suited for future use with a relatively small amount, if any, of manual input-output manipulations.

In the energy-handling field, coordinate systems for weapon launchers and their target-sighting equipment may be so chosen that a common mount is employed and the sighting equipment is merely displaced by the amount of the lead angle from weapon launchers. Thus the desired outputs are the line of sight to the target and the lead angle to the launcher. The proper selection of outputs and their associated coordinate systems can make quite a difference in the magnitude and nature of the systems problems that are incurred.

Another grouping of categories with which it is worthwhile to consider the outputs is the same as that used to describe the system inputs, namely, from their role as signals or information, power or energy or materials, and disturbances and noises. Attention will later be given to outputs from these points of view.

Ideal Outputs

A useful concept in formulating a description of the desired outputs is the so-called *ideal output*. One definition of the ideal output is that output which would be acceptably adequate from the point of view of the overall system requirements. As used here, "adequate" means that the output is satisfactory to meet the mission objectives without being either too utopian or yet too limiting so that the overall results are compromised.

What are the characteristics that this ideal output would have? How much power would it take to operate it, and what accelerations, speeds, positions, voltages, temperatures, pressures, bandwidth, repetition rate, switching speeds, etc., would each output have? How much in the way of performance, reliability, time, convenience, and cost is one willing to accept as satisfactory relative to the other quantities associated with the system? These individual outputs are in effect closely related to the objectives of the system, but they are more specific and describe each output of the system and its subsystem. They are not merely a broad listing of objectives.

Employing the ideal outputs as a base, one can now "build back" to establish what it is physically, or perhaps only conceptually, which will permit these outputs to be obtained. By the use of these secondary, physical or equipment requirements one is able to determine to what extent the real world as it presently is, or as it may reasonably be expected to be in the future, approximates that which is required to produce the desired outputs. What effects do the real-world requirements have on limiting the ideal outputs? Can these limitations be tolerated with changes in other parts of the system and its detailed requirements?

A modified definition of the "ideal" output in the Class A error sense as described below can be derived from this latter concept. Namely, the "ideal" output is that output which would be obtained from existing or from possible-to-attain equipment endeavoring to produce the desired output. Thus the ideal output as here defined includes Class A errors but is as good as one can reasonably expect with the power and materials equipment available, which has been as well mechanized or designed as possible.

Classes of Errors

A useful way of differentiating between the ideal outputs desired and the actual outputs obtained is to introduce the concept of classes of errors. Four different error classes can be used to identify the cause of the error or the deviation of the actual from the ideal.

Class A Errors. Differences between the ideal and actual outputs caused by inherent limitations in the particular method used for obtaining the actual output. These errors are introduced because of the system or equipment design selected and can be reduced or changed principally by design modifications.

Class B Errors. Differences between the output of a perfectly built equipment or system having Class A errors as designed and the actual output of a particular specimen as initially built. Class B errors are introduced because of the way this particular equipment or system has been manufactured, assembled, and checked out. Class B errors are established at the time of initial check-out and do not include any wear or operating-life degradation effects. Class B errors may be random or biased and may serve to increase or decrease the original Class A errors.

Class C Errors. Differences between the ideal output and the actual output of this particular system or equipment as initially built. Class C errors combine the effects of Class A and Class B errors and represent the net difference between what is desired ideally and what is actually obtained as first built. Whereas perfect design might reduce Class A errors to zero, and perfect manufacture and test might reduce Class B errors to zero, in reality there will be some of both types, and the resulting error seen in the system is the Class C error, which is not in general the sum of the Class A and Class B errors.

Class D Errors. Differences between the ideal output and the actual output of this particular system or equipment a period of time after it has been built and checked. Class D errors include the effect of degradation of output characteristics with time and usage. Whereas Class C and Class D errors are identical at the time of check-out, Class D errors will change with time, operation, maintenance, and service.

From the above consideration of the various classes of errors, it is apparent that there are a number of causes for discrepancies between the ideal output desired and the actual output experienced by an equipment or system in the field. For this reason it is important in the system formulation to keep in mind that all four classes of errors will exist in every system and not be carried away by the thought that if the inherent system design limitations producing Class A errors are reduced to zero a perfect (ideal) output will be obtained. It may be more realistic, less time consuming, and less costly in the long run to accept certain Class A errors initially and proceed with the production of a good overall system design that can be manufactured, installed, operated, and maintained.

Description of Outputs

The matrix of Figure 3.4-1 provides a way of categorizing the sort of information about system outputs that can be helpful in identifying and specifying these requirements. The classifications down the left side are the signal, power, and disturbances or noise headings that have been used in connection with the system inputs. The signal outputs and the power or material outputs in a sense represent the ideal or desired outputs. The disturbances or noise represent the unwanted or undesirable characteristics to be limited. Since these system outputs may serve as inputs to still other systems, it is reasonable that the same classification of inputs would be useful in describing the outputs.

The output use, definition, and description headings which go across the matrix serve to indicate first the application that is being made of the output, next a qualitative definition of the characteristics sought in it, and finally a quantitative description of their numerical values. Since the engineer must design his system to produce the outputs desired of it, it is essential that he know in as much detail as is required the complete characteristics of what the system is to put out. Man-machine relationships where significant must be spelled out in such a way as to be meaningful to the specifications of the outputs.

Signal Outputs. Output signals may be analog, digital, or hybrid and may provide an indication of the whole value or only incremental

Output classification \ Output Use Definition and Description	Application or Use Characteristics	Qualitative Definition	Quantitative Description
Signal outputs			
Power or material outputs			
Disturbances or noise			

Figure 3.4-1. Matrix of output classification versus definition and descriptions.

values. The application or use characteristics will first indicate what the output is to do and what kind of equipment characteristics it will have as its load. What quantity does the signal represent? Should it be analog, digital, etc.? Is the signal to be used for indication, display, or actuation? What sort of impedance or other interaction effects are to be expected from its load? Are there any special application considerations that should be known?

The qualitative definition of the output signal characteristics may incorporate the same terms as were used for input signals: gain, sensitivity, accuracy, deadband, saturation, linearity, etc., for analog terms; codes, repetition rates, signal equivalency, etc., for digital terms; as well as the other signal characteristics of dynamics, input impedances, and coordinate reference systems.

The quantitative definitions of the output signal characteristics provide numerical values for the various qualitative terms that have been given. Presumably not only nominal values should be given but also the tolerances which can be allowed.

Power or Material Outputs. Output power or materials must be described in terms of their nature as well as how they are to be used. Each specific output should be listed. Voltages to be used must be specified as direct current or alternating current and the power level given. Mechanical power, with speed and torque requirements, must be described. Chemical materials, including specific temperature, pressure, concentration, etc., must also be stated. It is important to identify the interaction of the characteristics of the use to which the output will be placed on the operation of the system which produces these outputs.

The magnitude, power level, regulation, dynamic response, reliability, and life must also be qualitatively defined and quantitatively described.

Disturbances or Noise. In the case of disturbances and/or noise the significance of the use or application term is that it serves to identify the particular output, whether it be of a signal, power, or material nature, to which the particular noise or disturbance applies. In a sense these disturbances and/or noise represent extraneous outputs which should be identified with the signal, power, and/or material quantities with which they are associated.

Although the signal and power outputs should have had associated with them their various classes of errors, if these have not already been identified this is an appropriate place for such data to be listed. For digital signal outputs, such information as quantization magnitudes and sampling rates, as well as wave-front characteristics and wave shapes, may be listed as a qualitative definition of the output disturbances and noise characteristics.

For analog signal outputs, null voltages, quadrature voltages, drift, positional or speed errors, and other such similar information are pertinent under the output disturbance and noise categories. Spikes and other unwanted signals caused by inductive and capacitive coupling with neighboring circuits and systems should also be identified. Side lobes, side bands, and extraneous harmonics in both space and frequency bands should be identified.

For power equipment, extraneous outputs may include voltage harmonics in time or space, acoustical noise, pulsations in power, voltage, or speed, impurities in chemicals, variations in temperature or pressure, regulation of mechanical and/or electrical supplies, and a number of similar unwanted conditions.

Although these disturbance and noise conditions are objectionable and should be consciously removed from the system if this is at all compatible with the constraints of time and cost, in most cases some of them do exist. It makes sense, therefore, to acknowledge their presence and to evaluate intelligently their effect on the system which these outputs are to supply. In this way the difficulties they may cause can be anticipated early enough to take appropriate steps to minimize their detrimental effects. When these steps are not taken early enough, the alternatives available to the system designer later may be much more limited, and the cost in time and money to introduce them may be considerably greater.

In summary, the purpose of system formulation is to recognize the magnitude of the job, including the pitfalls. Hence, both the formulation of the output requirements and an estimation of the difficulties caused by unwanted outputs are essential parts of a complete output description.

3.5 Problem Structuring and its Relation to Problem Formulation

The term structuring has a rather special meaning as applied to the system engineering problem. As used here,

Structure is the form, the arrangement of parts, the interrelationship of parts as dominated by the general character of the whole.

A related term is

Structured, having a definite structure, exhibiting organized structure or differentiation of parts.

As can be seen from these definitions, structuring is a way of subdividing or partitioning a large problem into a number of smaller problems which presumably can be handled separately and therefore more easily. Thus, structuring provides a method for reducing the apparent dimensionality of large system problems by decoupling the interaction between the different parts through the mechanism of organized separation.[41] Although this subdivision or structuring is performed to provide a first approximation to the solution of the whole problem, the interaction of the parts must be considered later once an apparently satisfactory first-order solution has been developed. The structuring assists in the more rapid realization of the first-order solution to the overall problem.

This view of the objective of structuring tends to place it in proper prospective as an important part of the methods of solving system problems. In the book *Operations Research and Systems Engineering*[21] by Flagle, Huggins, and Roy, under the section on problems of systems engineering, the problem of assembling a large number of diverse elements into an ordered whole is described and the iterative nature of the solution is highlighted.

"The process of engineering a new system consists of the solution of a series of problems involved in the assembly of a large number of diverse elements into an ordered whole. The solutions of these problems are basically different from those encountered in science. The terms of reference of each problem cannot be stated precisely; they are not solvable in closed form. Each problem has an infinity of solutions. The selection of the proper solution rests on judgment of the relative importance of incommensurate quantities. The problems must be solved by an iterative process of successive approximations. The parameters in the problems are time dependent. The final result is a compromise between the initially desired objectives and the capacity for realizing them with the manpower, time, and funds available. These are some of the problems characteristic of systems engineering." Furthermore, "It has been mentioned that one of the basic problems in system development is the complex interaction between the various components which make up a system. The technique which must be used to minimize the component interaction is to organize the components into subsystems in such a way that most of the interactions occur within the subsystems, while the individual subsystems interact as little as possible with one another."

Since systems engineering involves the bringing together of a number of equipments, parts, or subsystems which have some characteristic qualities such as inputs and outputs or boundaries of time or space, a number of possible structures of organization suggest themselves. Part of the problem of structuring is to reconcile the various possible structures and determine whether to shape the overall structure around one or more of the existing ones or to build the structure around one or more still different forms which may be more convenient to the overall system. This problem and others associated with structuring will be considered in succeeding sections.

Examples of Structuring

Although there will be more detailed examples of structuring in later sections, it is worthwhile now to present a few different types that can be useful.

Chronological or Time Structure. Figure 3.5-1 shows a block diagram in a simplified form of the system representing the industrial process about which we are concerned. As is signified by this figure, the research and development activities generally precede in time the design phase. Likewise the design precedes the production. In this one form of "differentiation by parts" these functional bases may have a strong and predictable orientation in a time sense.

Whereas for some industries, like automobiles, the time structuring described above is very well established, for other systems engineering jobs the same time relationship may not be so clear cut. Also, even for cases where the basic time structure is well established, a number of feedback paths may be introduced (see Figure 3.5-2) which because of their iterative nature serve to cause the general time structure to be modified. Although the existence of these feedback loops alters the detailed time relationships, it does not invalidate the simplifying

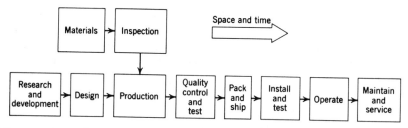

Figure 3.5-1. General industrial system structure, showing time-flow relationships between functions.

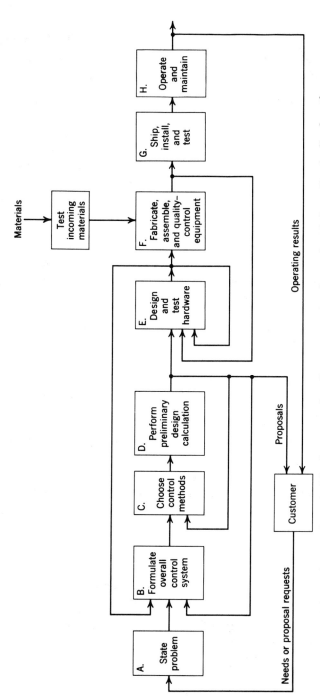

Figure 3.5-2. Automatic control business as a time-flow process, showing various information feedback paths.

structure concept that has proved most valuable in considering the time relationships involved in the industrial process.

Geographic or Location Structure. In connection with the study of the electrical power generation, transmission, and loads on a utility system, the geographic location of the principal loads and generating stations provides a meaningful structural form. Figure 3.5-3 shows in an abbreviated schematic form the fact that the major load and/or generating centers at A, B, and C are located physically at different points. Of course the loads and sources are not truly concentrated at three or however many other points are chosen, and the problem of what happens in the distributing of power within each of the locations is neglected in this simplified approach. Nevertheless, for such problems as whether to locate additional generation at A, B, or C, or whether to increase the transmission lines between A and B or A and C, the major structural form shown in Figure 3.5-3 is well suited. With this basic structure established, it is now possible to subdivide each of the major subsystems still further as may be required; see Figure 3.5-4.

In considering the control of temperature of a packed-bed reactor, the block diagram of Figure 3.5-5 indicates the major equipments, i.e., packed bed, furnace, controllers, and transmitters, which represent the process and its control. This is a system of reasonably manageable size, and the major parts of its structure are shown in their proper functional relationship to each other as well. With this as a start one can proceed to describe the dynamic characteristics of each portion, including power and energy capabilities and requirements, as well as the information which each provides and needs. The

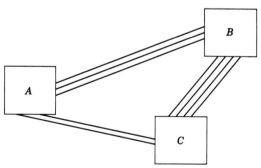

Figure 3.5-3. Geographic orientation of major electric utility load, generating stations, and transmission lines.

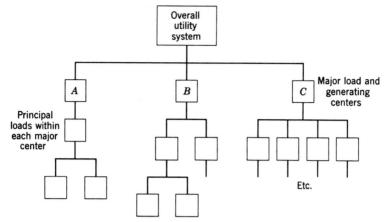

Figure 3.5-4. Further geographic subdivision of overall electric system of Figure 3.5-3 into finer geographic structure.

same basic structure is available to indicate these various forms of additional detail about this particular system. Since the various individuals and groups working on this system may have training and interests in some but not all of these different equipment portions, subdivision of the system in this way will generally be compatible with their particular abilities and responsibilities.

Functional Structuring. Although one normally associates functional structuring with the organization of equipment into parts—electrical, mechanical, aerodynamics, propulsion, instrumentation and

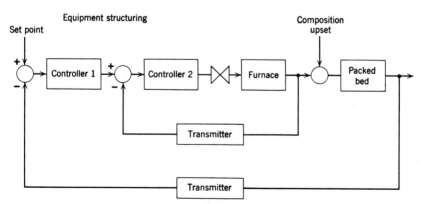

Figure 3.5-5. Block diagram structure of temperature control of packed-bed reactor.

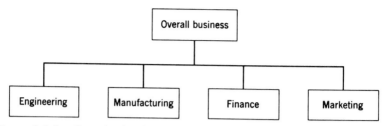

Figure 3.5-6. Organizational structuring of an overall business into major functional sections.

control, etc.—Figure 3.5-6 shows a different basis for structuring. Here an industrial company or business is divided into functional areas as considered by management, namely, engineering, manufacturing, finance, and marketing. Grouping the entire business activity into these four parts permits the assignment of personal responsibility to the individuals who head each functional area and provides a means of specialization of the training and talents of the people working in each area. Of course there are a number of activities which involve joint participation, planning, and action across these structural boundaries, but as a gross separation the functional division shown in Figure 3.5-6 represents a commonly employed method of structuring the organization of an industrial business.

The preceding examples of structuring are intended to be simply illustrative. They are not recommended as being either good or bad structural arrangements. They all depend rather strongly on modeling to illustrate their structural form, and certainly for this purpose modeling can provide a graphic as well as a more quantitative means of description. Later sections will provide additional detail about some of the significant structural methods and approaches in systems engineering.

Relationship of Structuring to Formulating

The approach one uses in solving a problem is greatly influenced by his understanding of it. Since the information and understanding gained in the formulation phase of the systems engineering process represent the starting point of the structuring effort for the solution of the problem, it is quite evident that the formulation of the problem will have a direct and strong influence on the system structure selected. However, the structure chosen must be one which can first be verified in an analytical fashion and then made to work in the real world. Thus, it must be capable of being checked back against the problem formulated and also against the real problem as it is known to exist.

Figure 3.0-1 illustrates in block diagram form how the jobs of formulating and structuring the system are interrelated. At the left of the figure, the information about the external problem serves as an input to the system design operation. It provides some general information in terms of objectives, requirements, environments, etc., without particular regard to the capability of the system. In all probability this information is neither complete nor accurate; as the formulating of the problem proceeds, one requests additional data from the source of information for the process being considered.

As a result of formulating, more detailed information is obtained about the problem, perhaps without detailed regard to the system capabilities. This initial formulation information serves as the basis for the initial structuring of the problem from which are generated two kinds of output information. One set of information is expressed in terms of what will be the results or outputs from the system. The other set is expressed in terms of the system characteristics necessary in order to obtain the results described in the first set.

With these two sets of information, the designer is now better armed with information about the system's characteristics and capabilities, which is fed back to both the formulating and the structuring portions of the effort. In addition to the possibilities for a modification in the problem formulation directly as a result of the system as structured, further requests for additional information about the problem may allow still more general information for formulation, resulting in a reformulated problem which now includes more of the capability of the system.

Likewise, restructuring may be required as a result of the initial structuring information about the system capabilities or the system characteristics or as a result of the revisions in the formulation of the problem. Several iterations may be required around the structuring-the-system loop by itself in terms of alternative structures which may include the nature of the equipment and the functional, as well as time and space, considerations involved. Although not specifically indicated on Figure 3.0-1, the intensive modeling, computing, synthesizing, optimizing, etc., necessary to accomplish the detailed systems analysis and synthesis required may be parts of the structuring effort. The necessity for using an iterative process of successive approximations of formulating and structuring is apparent from the foregoing.

Observations on Systems Structuring

Structuring introduces into systems a form of order and simplicity which appears to have many advantages. However, the advantages are not an unmitigated blessing, and a number of potential drawbacks to which one should be alert may be introduced by structuring. Furthermore, since some forms of structuring may be more efficacious than others for a particular system, it is worthwhile to try to establish the most favorable structure for the system under consideration.

First, structures tend to introduce interface problems at the boundaries between the parts of the structure. For example, simply by putting some electrical components in one black box and some in another, it becomes necessary to introduce two additional sets of electrical connections that can potentially go bad and cause a lower reliability for the system. Or by using a common power supply for a number of separate subsystems to minimize the possibility of having to regulate several different supplies to a common base, one may find that in order to operate one subsystem all of them must be energized. In addition, these interface boundaries tend to prevent certain otherwise logical interaction effects from taking place. For example, by generating two separate large quantities of power in different boxes and then subtracting them, one has less possibility of getting accurate data than if he generates the sum of the two large quantities and averages them, and at the same time in the same black box generates their difference.

Next, structures serve as models and provide for decision-making rules to be built-up around these models. There is a marked tendency for a system with sufficient energy and speed of response to try to conform to whatever reasonably good model exists for it. Hence, the choice of a particular structure tends to make certain results reasonably well assured, while for other structures other results are similarly certain.

Because the results obtained and the effort required to obtain them are frequently influenced by the form of structure employed, it is important to consider initially a number of alternative structures. In this way one is often able to pick out worthwhile points from several of these alternatives and obtain a few good structures for more intensive study in the detailed investigation.

Finally, when a well-organized structure exists, it tends to dominate the formulation process and hence to influence it in such a way that the structure most generally considered is that for which the prearranged structure is valid. Thus, there is merit in establishing a good

Functional Structuring

structure within which to work. With such a structure the formulation procedure can be handled readily and well. With a poor structure or no structure, formulation is difficult, as are other portions of the overall systems engineering procedure.

3.6 Functional Structuring

From a pessimistic viewpoint, it can be stated that there is no good general way of structuring a system. However, from an optimistic point of view one can say that a number of good ways of structuring systems exist and that some are better than others for any particular system. In this and the following sections, there will be a presentation of a number of structuring approaches that have merit and have been employed successfully, including functional structuring, equipment structuring, and use of various coordinate systems. The purpose in considering these alternative methods is to expose a number of different features of structuring that may be useful in certain systems engineering problems. Their variety should serve to emphasize the facts that a large number of structuring schemes are possible and that a person should be willing to consider a number of structures in order to find one or more that will best serve his needs.

Structuring Concepts

As has been pointed out earlier, structuring is a way of giving form or order to the large number of parts that comprise the system. Since there are numerous forms that the system may take, it is worthwhile at this time to point out that many different structural arrangements may be possible and that in the formative stages of structuring the system it is desirable to consider a number of these arrangements.

As a simple illustration of this point, consider the area shown in Figure 3.6-1. As a means of describing a particular part of that area it is possible to apply a number of different schemes of subdivision, as shown in Figure 3.6-2.

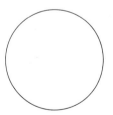

Figure 3.6-1. Circular area represented as whole system.

Parts A-E emphasize the area as a whole, whereas part F stresses the fact that the whole may be made up of a number of unrelated

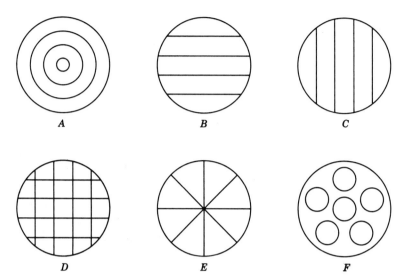

Figure 3.6-2. Circular area structured into various subareas.

parts. Part *A* tends to accentuate the circular nature of the outline of the area, as does to a certain extent part *E*. On the other hand, parts *B* and *C* tend to de-emphasize the character of the outline and bring out a vertical or amplitude stratification (part *B*) or a horizontal or sequentional stratification (part *C*). Part *D* with its subdivision into squares tends to give a much finer granularity or mesh to the entire area and to reduce the significance of the natural curvature which may have characterized the original area.

It is interesting to note that many of the curved potential-field type coordinates in nature are in sharp contrast with the straight lines and circles employed by man-made structures. This is not to imply that one or the other is always better but merely to observe that man has been able to live successfully with both and to use each to his advantage.

Structuring in some ways appears to be a method of providing something akin to the idea of "invariance" used in control systems. By employing the invariance principle one finds what characteristic(s) in the particular problem or problem-type being considered tends to remain unchanging or slowly changing (invariant) with time, so that one can predict the future with some degree of assurance based upon the past or present observations. Hopefully the structure provides this framework within which information about one part of the system provides some sort of a priori knowledge about the remainder. In some cases this knowledge is simply the assurance that this part can be changed without affecting appreciably the operation of other parts. In other cases knowledge of what a set of values is in one equipment means that the corresponding set of values in another equipment can be directly expressed. Unfortunately the choice of structures does not always lead to such simple or direct relationships between portions of a system.

It is of interest to note that the use of more than one structural arrangement for a given system may be convenient. Thus, for the purpose of setting up a time structure of a system, a chronological arrangement may be most appropriate. From the point of view of hardware to be built, an equipment structure may be most useful. For some purposes, it may be highly desirable to have available a combination of both chronological and equipment structures. Generally speaking, the system designer is not limited to one structure; however, it is worthwhile to try to limit the number of structures to a reasonable value, say six or less.

Structuring, although it can provide convenience and simplicity, also constitutes a possible source for misunderstanding and confusion. Not all people understand the system in the same way, nor do they see the relationship between one part of the system and the other portions in the same way. Such differences may be a function of the prior training, experience, or frame of reference of the individuals or may be attributable to other reasons. The structural relationships should be ones which are reasonably clearly defined between each part and every other part. In this way it should be possible to identify clearly which part of the structure one is dealing with and hopefully to express the rules or conditions which apply for that part. The process of selection of the structural form should be considered as having potential pitfalls and difficulties, and the structuring should be reviewed with the responsible people involved to be reasonably sure that the structure ultimately selected can be agreed upon by a substantial majority of those directly concerned.

Viewed in another light, structuring can be said to provide a way of accentuating the similarities and/or differences between the various parts of the system. To the extent that the similarities and differences are real and apparent to all, the structure is convenient. To the extent that this situation does not exist, the structure provides a possible cause of difficulty and confusion in its use.

By bringing out into the open early enough the relationships that may cause ambiguity and confusion, the structure performs a useful function because it tends to force the persons involved to come up with a better understanding of the system and a more precise structuring of it.

Viewed in still another light, structuring lends itself to providing a basis for modeling of the system. Structuring tends to be a coarse analog form of modeling which can later be reworked in more detailed form to describe more qualitatively and quantitatively the basic as well as the specific forms of the structure.

Emphasis on Physical Needs and Limitations

The term functional structuring is used to describe the approach to structuring in which the functions required to accomplish the objectives of the system provide the basis for its structure. The physical laws describing the energy, materials, or information relationships are the starting point of this structuring method, and frequently the system is built around some inherently simple analogy of one or more basic physical laws. An orderly set of relationships is sought which will provide the performance or operation required or desired when the approach is understood and found to be workable and satisfactory; then the equipment necessary to implement it can be established as a secondary step. The primary step is to build the initial order or organization around the requirements of the system.

Start with Ideal or Hypothetical Operation. The starting point for the problem solution is to deal with a fundamental or so-called "coarse" approach in which an idealized or hypothetical representation of the overall problem is considered in a simplified form. Assuming perfect (or ideal) operation of the equipment, with certain approximate maximum (or minimum) forces or torques, velocities, temperatures, pressures, flows, voltages, etc., in what way will the most favorable operation occur? Starting with a given set of requirements, it is possible to determine in what way and with what parameter values the desired objectives can be met. The most critical needs tend to influence the selection of methods for accomplishing the desired results. In general, less significant factors are considered only approximately at first, if at all. Later, after the first-order effects are determined, the influence of these major factors on the extent to which second-order effects must be included is considered.

For example, having established the output power required of a motor, one than determines the effect of the motor itself on the overall size and input power rating of the motor required to handle both itself and its load. Thus, if the power of the load requires a given

Functional Structuring

torque and speed,

$$\text{Power} = T \times S$$

the structure of the system may be built around the torque- and speed-producing characteristics of the motor; the speed-, torque-, and/or power-measurement capabilities of the instrumentation system; and the control means for regulating the power input to the motor. The motor-size requirements to supply the desired shaft load thus determined must of course include not only the load needs but also the needs of the motor itself when operating under the load duty cycle.

Consider a radar range-power problem as another illustration. Having established an ideal power-area product relationship to obtain a given radar detection range, one next determines the actual radiation area necessary to obtain the desired ideal area. The emphasis is placed on what conditions or functions are needed to accomplish the system objectives; the method of accomplishing these functions is then based on obtaining means to provide what the functions require.

Thus, if the various system requirements, R's, can be stated in analytical form,

$$R_1 = f_1(A_1, B_1, M_1),$$

and

$$R_2 = f_2(A_2, B_2, D_2, X_2),$$

and
(3.6-1)

$$R_3 = f_3(B_3, D_3, X_3, Y_3),$$

and

$$R_4 = f_4(A_4, M_4, Y_4)$$

the structure will tend to be built around the variables or functions, A, B, D, M, X, and Y. Means for accomplishing these functions or portions thereof become the objectives of the system design, and the efforts and energies of design and component groups are directed at achieving these various desired functions.[25]

Perform Sensitivity Analysis to Determine Criticality of Function or Parameters. Although the ideal operation or the mathematical equations provide a good start at system structuring, it is essential to consider how sensitive the results are to the actual numerical values of the various parameters and relationships. Partial derivatives of the various results with respect to the different parameters indicate how critical the results are to parameter changes, and therefore the extent to which it is necessary to monitor or control these parameters.

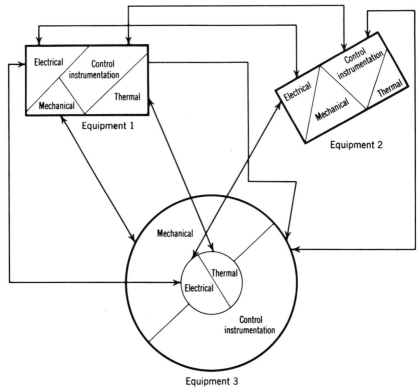

Figure 3.6-3. Interactions of equipments and functions, stressing equipment considerations.

Thus from Equations 3.6-1, one can determine

$$\frac{\partial R\text{'s}}{\partial A\text{'s}}, \frac{\partial R\text{'s}}{\partial B\text{'s}}, \frac{\partial R\text{'s}}{\partial D\text{'s}}, \quad \text{etc.} \qquad (3.6\text{-}2)$$

from which the amounts of change to be realized from changes in the parameters are evaluated. When these changes in the results with changes in parameters are unacceptably great, it may be essential to modify the system organization or means necessary to accomplish the desired results, as, for example, by limiting the extent of the changes in the parameters that were found objectionable. This in turn calls attention to the fact that additional parts of the system structure may be required to provide this limitation of the parameter variation which has been found necessary. As an illustration, consider the case in which temperature variations of a gyroscope rotor and pick-off may cause errors proportional to the changes in tempera-

Functional Structuring

ture. By adding a temperature control to the housing of the gyroscope the range of rate errors from the gyroscope may be greatly reduced. Thus the addition of the temperature control to the basic elements making up the overall gyroscope assembly may improve the resultant performance appreciably.

Structuring by Equipment and by Technical Function

Frequently in modern systems a number of different equipments each having several functions must operate together in a compatible fashion. One way of structuring such a system is on an equipment basis. As shown in Figure 3.6-3, each of the three equipments that make up a system may have in turn four or more separate functions to perform, namely, mechanical, electrical, thermal, and control instrumentation. Although within each equipment there are interactions of each function with each other, interactions also occur between equipments of corresponding electrical, control instrumentation, mechanical, and even thermal functions in the fashion shown by the arrows.

One might assign the responsibility for each equipment to a separate individual and thereby structure the system on the basis of equipments alone. An alternative method is to assign responsibility in such a way that each of the four functional area is handled in a consistent fashion regardless of what equipment it is a part. Figure 3.6-4 shows

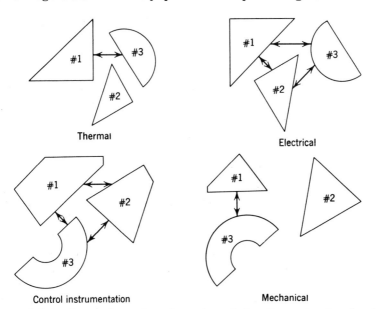

Figure 3.6-4. Interactions of equipments and functions, stressing function considerations.

	Function			
Equipment	Electrical	Mechanical	Thermal	Control Instrumentation
#1	X	X	X	X
#2	X	X	X	X
#3	X	X	X	X

Figure 3.6-5. Equipment-function matrix in coarse form.

how the four different equipment parts can be grouped together and how the compatibility of the methods of handling each function regardless of the equipment with which it is associated can be evaluated. Furthermore, the nature of the functional interactions between equipments is also emphasized in this fashion and is thereby more likely to receive proper attention. This addition of the functional structuring can be of great help in providing greater overall capability to the system, including both its operation and its method of manufacture. Figure 3.6-5 illustrates how the combination of equipment and functional structuring may be represented in a matrix form. An alternative matrix form, which stresses the interdependence by functions across equipments and their interfaces, is shown in Figure 3.6-6, where an X indicates that a function in the equipments listed vertically affects the functions listed horizontally. Although the structuring concept may tend to separate the system into equipments and functions, it is essential to realize that the system itself will have its own functional interrelationships which must be recognized and duly accounted for in order for the system to perform satisfactorily.

Structuring by Time or Operational Phases

Because of the tendency for changes to be required in systems with time, there is a natural evolution of systems that may take place in an unorganized fashion. However, it may be possible to structure an orderly evolution so that it is more readily understandable and useful to the many people involved with operating, servicing, or maintaining the system. Generally an appreciable advantage is gained if the system is engineered in such a way that it is divided into distinctive time periods with significant characteristics associated with them. One obvious time characteristic is differentiated by a time period, e.g., the annual car-model change. Another time characteristic is the time when a certain function in the production of the system has been completed, such as the end of development or the beginning of design. Still a third time distinction is associated with the different phases of the operation of the actual system, i.e., the end of warm-up or the completion of the launch phase. By identifying these time periods and structuring the system description and operation around them, it may be possible to decouple the various activities in a more workable fashion.

Time Established by Equipment Requirements. Changes in the equipment making up a system may be required to incorporate new materials, methods, or performance. To structure these changes in a convenient and understandable fashion, one can develop a schedule indicating what changes should be introduced and when these will occur. Figure 3.6-7 shows such a schedule. It indicates that the original design, consisting of equipments A, B, and C, is to be used for two years. However, after one year, the Mod I design will be introduced, using equipments B and C of the original design but replacing A and adding D. Because of the need for introducing Mod

	EQUIPMENT	#1				#2				#3			
		Electrical	Mechanical	Instrumentation	Thermal	Electrical	Mechanical	Instrumentation	Thermal	Electrical	Mechanical	Instrumentation	Thermal
#1	Electrical	X	X	X	X	X				X			
	Mechanical	X	X	X	X						X		
	Instrumentation	X	X	X	X			X				X	
	Thermal	X	X	X	X								X
#2	Electrical	X				X	X	X	X	X			
	Mechanical					X	X	X	X				
	Instrumentation			X		X	X	X	X			X	
	Thermal					X	X	X	X				
#3	Electrical	X				X				X	X	X	X
	Mechanical		X							X	X	X	X
	Instrumentation			X				X		X	X	X	X
	Thermal				X					X	X	X	X

Figure 3.6-6. Equipment-function matrix in intermediate form.

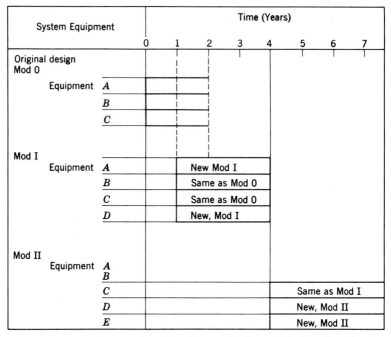

Figure 3.6-7. Structuring of equipment in time Mods (modifications).

I soon as well as the need for continuing production of the original design, there is a year's overlap of the two models.

After four years, there will be a Mod II, which eliminates equipments A and B and includes original equipment C with new equipments D and E. Because there is enough time to plan for the changeover from Mod I to Mod II, the schedule calls for the production of Mod I to cease at the same time that Mod II is begun.

Time Established by System Function Requirement. In planning the work functions of a system, one develops a structure of the different jobs to be done. The preparation of a graphic schedule of work functions versus time provides a time structure by which the progress of the job can be judged from a technical-accomplishments point of view. Figure 3.6-8 shows such a schedule for an automatic control business; the functional description is given at the left, and the corresponding objectives to be accomplished are tabulated at the right. These functions are shown versus time and indicate that a number of them take place concurrently.

By virtue of having a functional structure of work, as in Figure

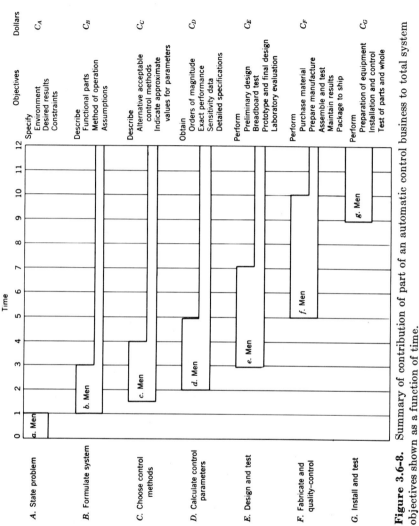

Figure 3.6-8. Summary of contribution of part of an automatic control business to total system objectives shown as a function of time.

3.6-8, it is possible to identify particular times at which a review of the work status in each function is most pertinent. Thus at time 1, emphasis can be placed on the status of the problem statement and on the start of formulation of the system. At time 1½ presumably not a regular review period, it is nevertheless appropriate to determine whether the choosing of control methods has been started. Depending on the original choice of the structure or division of the work portions into the seven parts shown, one arrives at different times for review and different topical emphases for the reviewing. It is apparent that the type of functional structure chosen can be of appreciable significance to the development of a system.

Time Established by System Operational Phases. Many systems have a number of modes or phases of operation during some of which equipment or subsystems operate in one fashion whereas during others they operate differently. Because of these different operational phases, an important time structural configuration is that built around these equipment-mission phases relationships. Figure 3.6-9 shows such a matrix and indicates the major vehicle equipment stages to

Space Vehicle Stages	A. Prelaunch	B. Launch	C. Earth parking orbit	D. Injection into earth-moon trajectory	E. Lunar parking orbit	F. Lunar descent	G. Lunar launch	H. Lunar rendezvous	I. Injection into earth-moon trajectory	J. Re-entry	K. Recovery
1. S-IC		X									
2. S-II		X									
3. S-IVB		X	X	X							
4. Instrumentation unit		X	X	X							
5. Lunar excursion module			X	X	X	X	X	X			
6. Service module			X	X	X	X	X	X	X		
7. Command module			X	X	X	X	X	X	X	X	X
8. Ground operational support systems			X	X	X	X	X	X	X	X	X

Figure 3.6-9. Space-vehicle stages versus mission phases matrix.

Functional Structuring

the left and the mission phases when they are operative. As is shown, some stages are operative for only part of the flight while others are operative throughout.

Another form of similar equipment-operation matrices could be built around various alternative modes of operation that might occur as a result of one or more abnormal operating conditions corresponding to different situations that might arise during the lifetime of the system. Although many of these operational phases may never occur, or must be avoided if at all possible, structuring these time phases as situations to be considered is essential.

Structuring of Organizational Work Groups

A perennial problem of any organization which is involved in systems work of a changing nature is the need for the organization itself to change as it adapts to new opportunities, new objectives, new equipment, and new people.[26] Depending on the structure chosen for the organization, there may be a significant change in the ease or difficulty of operating and the nature of the results that can be achieved. The purpose of this material is merely to indicate briefly a number of alternative organizational structures that might be considered for an engineering laboratory group. Depending on the size, method of financing, objectives, etc., there are advantages and disadvantages for each of the various organizational structures described.

Marketing or Consumer Orientation

>Defense
>Industrial
>Utility
>Consumer

This type of structure emphasizes the use of the product of the engineering effort and stresses the ability of the organization to cater to the specific markets to be supplied and the operational environments peculiar to each. However, it may result in similar technical activities being performed in each of the four parts of the structure.

Product Focus

>Motors, generators
>Transformers, reactors
>Lighting products
>Communication products
>Appliances
>Transportation equipment

This structure stresses existing products and tends to have certain short-time advantages. Although it may concentrate some people with kindred technological interests together, this is not a major objective of the product structuring method.

Scientific Orientation

    Mechanics
    Physics
    Chemistry
    Metallurgy

A scientific orientation lends itself to the development of technical specialists with strong allegiance to the traditional university training efforts. Of itself, such an organization is not likely to be product or applications directed, although it may make significant original contributions.

Engineering Fundamentals Orientation

    Energy
    Materials
    Information

These three engineering fundamentals are the ones which we have noted earlier as broadly encompassing the field of systems engineering. As such they cut across the scientific fields and have more of a user emphasis. Again they lend themselves to consultative and research types of action and require the addition of application-type situations presenting the important elements of realism and urgency in order to bring the ideas to the marketplace.

Project Plus Engineering Fundamentals

    Energy
    Materials
    Information
    Projects

This structure builds upon the advantages of the previous one of engineering fundamentals but in addition stresses applications through its project efforts. By providing a means for personnel to be able to work on projects or in one of the engineering-fundamentals areas, as is required, this structure lends a measure of flexibility that is frequently desirable in engineering organizations. However, in some respects the continual shifting of people and objectives may introduce some difficult situations within which to operate. It will be noted that this organization arrangement is similar to that described in Chapter 2.

The influence of the form of structuring on the results that can be obtained may be appreciable. By considering first a coarse approach to a number of different forms of structures and then later a more detailed or fine approach for those which appear to have merit, one is able to bring an overall order to system organization that can be well worth the effort spent.

3.7 Equipment Structuring

The engineering of systems in industry is vitally concerned with the design and use of equipment, since hardware is basically what a large segment of industry makes and sells. In *Systems Engineering Tools*[49] the point was stressed that this equipment may be used for the same or different customers at the same or different times. In this section are discussed the approaches used in conceiving and organizing a system plan for structuring an equipment design so that this interchangability, with its accompanying savings in cost, time, and utility, is achieved. Likewise discussed are general ways in which long-term (10–40 year) evolutionary changes in systems may take place so that judgments affecting short-term (1–5 year) decisions can be made with both long- and short-range considerations in mind. Also to be presented in this section are problems associated with a system which initially is so novel that the primary concerns are to provide a way for this unique system to work, without worrying too much about how best it should be made a part of a broad line of such systems.

Long-Term Equipment Evolution

A. D. Hall[34] points out that two basically different approaches to system design are available to the engineer. One approach involves the idea that a given system is essentially customed designed, so that it must be considered as a *whole*. "Every part of the system is so related to every other part that a change in a particular part causes a change in all other parts and in the total system." A system exhibiting such wholeness is optimized on the basis of its own needs, and no particular thought is given to interchangeability of parts with other systems or use of a standard line of equipment.

The second approach involves the idea that a system is so designed of independent parts that each part may be replaced and the action of the remaining parts be relatively unaffected; hence the name *independence*. Each set of parts which makes up the system is completely unrelated to the remaining parts. A change in any part depends entirely on that part. The variation in the total system is the physical sum of the variations of its parts. A system exhibiting a fair measure of independence is optimized on the basis of optimizing each of its parts. The result of suboptimization, that is, separate optimization of each of the subsystem parts, should produce a total system which is also optimum. A system composed entirely of independent equipment is said to be a *factorable system* because it may be subdivided into separate parts each capable of operating independently of each other. A system composed of standard parts, capable of being used compatibly with a number of other standard equipments, exhibits a high measure of system independence.

Actually, as Hall points out in his book *A Methodology for Systems Engineering*,[34] wholeness and independence are not separate properties but are extremes of the same property. The following few paragraphs borrow heavily from his work. Wholeness and independence are matters of degree, but no easy way seems available for using a scale to represent the measure of the property of degree. Although the properties of independence and wholeness are difficult to evaluate on an absolute basis, two terms, *progressive factorization* and *progressive systemization*, are used to describe trends toward independence and toward wholeness, respectively.

Progressive Factorization. Progressive factorization corresponds to a condition of growth in which as the system gets bigger there is

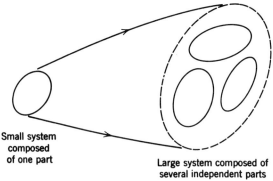

Figure 3.7-1. Progressive factorization as a small system grows into a large, partitioned one.

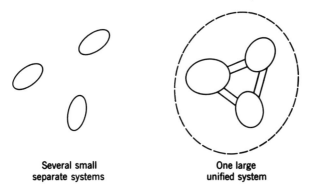

Figure 3.7-2. Growth of small independent systems into large unified system.

a tendency for an increasing division into subsystems and subsubsystems, which in their new size are sufficiently large to be handled almost independently. For a small business, the owner, manager, and worker may be only one person. When the business is so large that thousands of people are involved, it may be divided into several different divisions, each of which in turn is made up of different functional groups like manufacturing, engineering, marketing, and finance. Figure 3.7-1 shows conceptually how a small system growing into a large one is subdivided as it experiences progressive factorization.

Progressive Systematization. Progressive systematization represents the condition of growth in which as the system gets bigger there is a tendency for the separate parts to grow together and become one whole. This cohesion may come about through the strengthening of pre-existing relations among the parts, the development of relations among parts previously unrelated, the gradual addition of parts and relations to a system, or some combination of these changes. As an example, consider the development of the long-distance telephone network. First local telephone exchanges sprang up about the country. Then exchanges were joined with trunk lines. As transmission techniques improved, more exchanges were added at greater distances. Later, distance dialing was introduced, placing the network at the command of operators and eventually of customers. The record has been one of increasing unification of the system. Figure 3.7-2 shows conceptually how several small separate systems growing together into one large one are unified as they experience progressive systematization.

Hall points out that it is possible for progressive factorization and systematization to occur in the same system either simultaneously or sequentially. Consider the early history of America during which groups of people colonized various parts of the country. These groups became more and more independent of their parent countries. Gradually, the new country became unified as further interchanges occurred between the groups, common traditions were established, and a new government was formed.

Centralization. Another idea which is of significance in connection with system evolution is the centralization concept. A centralized system is one in which a single subsystem or part plays a major or dominant role in the operation of the system. We may call this the leading part or say that the system is centered around this part. A small change in the leading part will then be reflected throughout the system, causing considerable change.

Either progressive factorization or progressive systematization may be accompanied by progressive centralization; as the system evolves, one part emerges as a central and controlling agency. The concept of the centralized system yields the important principle that the more a system is centralized, the more the leading part must be protected against damage from uncontrolled environmental factors. When the risk involved in keeping the leading part centralized is so great as to outweigh the advantages of the centralization, a trend toward decentralization may occur. Then several factorized subsystems are given leading parts which are centrally monitored only in part or not at all.

The foregoing material sheds light on the general problem of long-term equipment evolution in that it indicates a number of long-term evolutionary and perhaps contradictory trends to be possible. It emphasizes the fact that the needs of the system must be determined in a quantitative as well as a qualitative fashion before sound judgment is possible regarding the proper method of structuring for the current situation.

Equipment Lines as an Aid to Structuring

In considering equipment lines relative to the subject of structuring, one is placing principal emphasis on the cost, time, and reliability objectives of the system rather than specifically on the performance goals. Presumably, the products involved, such as refrigerators or motors, are already relatively well known as to their principles of design, manufacture, and operation. The problem is one of having designs which are more compatible from model to model, from component to component, and from year to year. In this way the initial costs may be spread out over a greater number of units. The time required to develop a new design or to make the equipment on an incremental basis is reduced. Also, since large-scale, repetitive manufacturing, assembly, and test procedures can be standardized, the equipment manufactured can be more uniform and reliable.

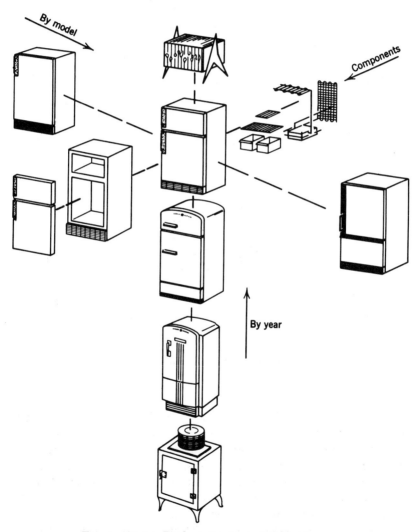

Figure 3.7-3. Product structure—refrigerators.

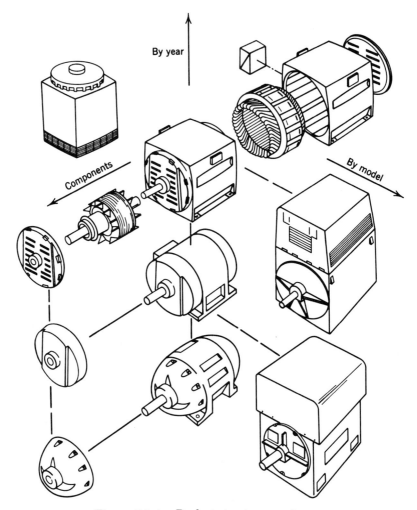

Figure 3.7-4. Product structure—motors.

Refrigerators and Motors. To illustrate further the nature of what is meant by the product structure of equipment lines, Figures **3.7-3** and **3.7-4** from D. F. Langenwalter show generically the product structure for refrigerators and motors. These diagrams indicate that there are at least three structuring arrangements along which a product may be considered: year, model, and component. The first of these, by year, is shown on the vertical axis.

In the case of the refrigerator (Figure **3.7-3**) one notes from the different exterior shapes located one above the other how the old monitor-top refrigerator through the years has changed in overall appearance. Furthermore, during any one year, as shown by the diagonal line from upper left to middle right, a number of different models are available for sale and are all being made, stocked, and maintained. In addition, for any year and model, there are a large number of components, as shown by the variety on the axis from upper right to middle left.

Figure 3.7-4, showing the product structure for a motor line, has more of the structure indicated. The basic year by year motors are not only shown on the central vertical axes, but also indicated by vertical displacement of the different motor end shields as well as whole alternative motor models. Again in this case the diagram shows a number of the various components making up the model for a given year.

Each equipment model has certain basic components and then also other features or accessories which may or may not be present in other models. Furthermore, with product structure pictures like these, the engineers and designers can appreciate the possibilities of standardization, which will make the various models easier to manufacture in a given shop while still meeting the marketing requirements for variety and change. A good product structure, also makes possible an orderly introduction of new features with a better knowledge of their impact on the design and on the shop.

Automatic Control Equipment. A starting point for structuring an automatic control equipment line is an appreciation of the relatively few functional tools which can be used to describe most automatic control systems. Nine such functional tools are *sensing, converting, storing, communicating, computing, programming, regulating, actuating,* and *presenting*. Table 3.7-1 contains definitions for each of these terms.

The functions of sensing, actuating, and displaying are ones in which the control equipment is quite directly coupled to the external or internal inputs, the process manipulators, or the presenting outputs; see Figure 3.7-5. Hence, it is quite likely that these equipments, which serve as inputs or outputs for the remainder of the control, will vary with the customer and his process needs. The speeds and accuracies of the automatic control equipment will be required to be comparable to and/or significantly better than those required by the process and the operator.

Although the functional tools of automatic control have been shown

Table 3.7-1. *Table of Definitions of Information-Conversion Functional Terms with Respect to Equipment Application*

Function	Definition or Explanation
Sensing	Generates primary data which describe phenomena or things.
Converting	Changes data from one form to another to facilitate their transmission, storage, or manipulation.
Storing	Memorizes for short or long periods of time data, instructions, or programs.
Communicating	Receives data and transmits it from one place to another.
Computing	Performs basic and more involved mathematical processes of comparing, adding, subtracting, multiplying, dividing, integrating, etc.
Programing	Schedules and directs an operation in accord with an overall plan.
Regulating	Operates on final control elements of a process to maintain its controlled variable in accord with a reference quantity.
Actuating	Initiates, interrupts, or varies the transmission of power for purposes of controlling "energy conversion" or "materials" conversion processes.
Presenting	Displays data in a form useful for human intelligence.

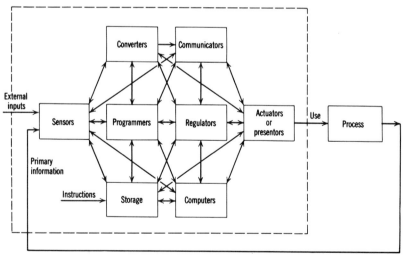

Figure 3.7-5. Block diagram of functional tools of automatic control connected to process.

Equipment Structuring

to be few in number, the actual hardware required to accomplish the various tasks can be myriad in nature. Specifically, such practical considerations as speed of operation, voltage levels, impedances, complexity, ruggedness from both an electrical and a mechanical point of view, and ambient temperature will vary from application to application. One approach to this problem would be to design each equipment for each application as required, but this would be time consuming and expensive. Another approach would be to design in such a way as to have one equipment meet all the limiting conditions and have all the items available in one design. The latter procedure would by no means represent an economical solution because all equipment would require the most expensive components.

An alternative design organization approach is to divide the overall hardware configurations into a relatively few ranges of operation which are characteristic of the major areas of application for automatic control equipment. The structural organization will be built around the needs for the entire information conversion job.

The following four ranges have been suggested by the Brown Boveri Co. as most suitable:

Range 1: Power Engineering Application—for open-loop control, analog closed-loop control, and automatic control in industry and utilities. The main requirements are concentrated on high mechanical strength and insensitivity to external electrical disturbances. The power involved is relatively high, and the working speed low. The control units have only to perform fairly simple operations and to make quite simple logical decisions.

Range 2: Telemetry—for reliable transmission and reception of a large number of control commands of all kinds, position signals, and measurements, using only one conductor. This range covers those cases in which the individual parts of an installation are so widely separated that the exchange of information between various places can take place over a simple communication link. Compared with the requirements which have to be fulfilled by Range 1, the conditions regarding ruggedness are somewhat reduced. The power level is much lower but the operating speed is much higher than in Range 1.

The main task of the control units is coding and decoding of commands and incoming formation, testing for undisturbed transmission, and the introduction of a time graduation for signals which may possibly have been received simultaneously.

Range 3: Counting Applications—for rapid controls and for the digital and automatic control of machine movements. In regard to ruggedness, power level, and operating speed the conditions correspond roughly to those in Range 2. The main task of the control units is to count and evaluate periodic phenomena.

Units of Ranges 1 and 3 can be employed without modification to solve certain tasks involved in the protection of power and industrial systems. This is particularly true of complicated protective duties in which calculating operations have to be performed.

Range 4: Digital Computing Applications—for tasks in which complicated logical decisions have to be made at very great speed, and arithmetical operations performed with great accuracy. Typical applications are digital control systems permitting deviations from the desired value of less than 0.01 per cent and arrangements enabling the optimum performance to be obtained with a particular process. The power level corresponds approximately to that of Range 2; the ruggedness is slightly reduced, but the operating speed is more than a whole order of magnitude higher.

Table 3.7-2 lists a number of significant electrical characteristics for each of the four ranges listed above. The values given are ones that are being used by one manufacturer of a broad range of automatic control equipment. They cover the nominal frequency range, the magnitude and rate of change of voltage with time, the approximate impedance levels, and the magnitude and tolerances of the supply voltages. In effect they provide a structural form for the application, design, and manufacture of automatic control equipment.

In addition to the importance of an electrical hardware structural compatibility, there is also a need for mechanical design structural compatibility for units in each range, as well as for units in different ranges.

Selection and Grouping of Equipment

In connection with the structuring of equipment two recurring questions relative to the selection and grouping of equipment warrant brief mention here, although the most appropriate answer may differ depending on the characteristics of the particular system. One such question is whether the equipment should be general-purpose (standard) equipment adapted to the needs of the particular system or special-purpose equipment designed and manufactured especially for the system. Of course there are intermediate positions between these extremes which will allow a certain amount of general-purpose equipment with the remainder to be special-purpose equipment.

Table 3.7-2. *Significant Electrical Characteristics for Different Application Requirements as Noted*

Performance Characteristics				Range			
				1 Power Engineering	2 Telemetry	3 Counting	4 Digital Computing
Maximum repetition frequency				1 kc/s	20 kc/s	100 kc/s	1 Mc/s
Standard signals	Analog			±15 v	...	...	...
	Digital	Code	0	0 to −1.5 v	Approx. −1.5 v	+0.2 to −0.6 v	Approx. −0.5 v
			1	−15.2 to −24 v	Approx. −5 v	−4.2 to −6 v	Approx. −5 v
		Wave front		...	2–4 v/1 µs	Approx. 4 v/0.4 µs	1–2 v/0.3 µs
Approximate impedance values		Output		0.2/1 k ohm	1 k ohm	1 k ohm	1 k ohm
		Input		4/22 k ohm	Dynamic	4.7/22 k ohm	Dynamic
Supply voltages and tolerances				+24 v* + 6 v† ± 2% 0 v	+6 v 0 v ±30%	+6 v 0 v	+1.5 v 0 v

* Analog units.
† Digital units.

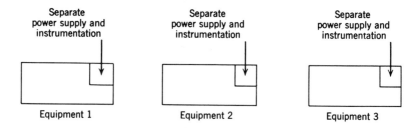

(a) Separate Power Supplies and Instrumentation

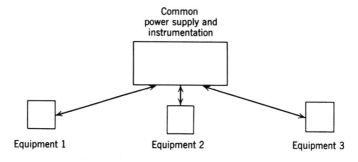

(b) Common Power Supplies and Instrumentation

Figure 3.7-6. Comparison of separate versus common power supply and instrumentation configurations.

The second question relates to the matter of whether such leading systems are power supplies, control instrumentation, and computers should be centralized or separate and associated with each of the major equipments with which they have a strong functional tie. Figure 3.7-6 shows in block diagram form the nature of the problem. For separate power supply and instrumentation each equipment can operate independently, and failure of one set of power supply or instrumentation does not prevent the remaining units from performing adequately. With a common supply of power and instrumentation, more accurate regulation, a better load factor for the power supply, and better overall instrumentation accuracy may be realized.

It is apparent that an appreciable difference in the equipment structuring may result, depending on whether central or separate power supplies and instrumentation are employed.

Equipment Interface Problems

Another equipment structuring problem of importance in systems engineering is related to the selection of the interfaces of the system.

Conclusions

Figure 3.7-7. Choice of common equipment versus separate equipment as it affects interface problems.

As mentioned earlier in this chapter, the selection of a stystem structure with its associated boundaries introduces the need for electrical and other interconnections externally between equipments that with another system structure would be internal connections between subsystems of the same equipment. Since the reliability of interequipment connection between subsystems may be much less than that of intraequipment connections between the same subsystems, the choice of equipment systems structure can have an important bearing on overall reliability. Figure 3.7-7 illustrates this interface problem.

In general, it is desirable to raise low-level electrical signals within a given equipment before sending them to another external equipment. Of course digital methods tend to increase the tolerance of the signals to noise and other variations in parameters over analog techniques so that the interface problems may be reduced somewhat in this fashion. The interface problem is one of real significance which can be improved or aggravated by the choice made in system equipment structuring.

3.8 Conclusions

The operations of formulating and structuring a system are very important to its overall success. However, the processes of formulation and structuring may differ in detail from system to system, even though the general procedure of bringing good understanding to what may be initially a poorly defined problem is required in each case.

There are a number of basic principles which are applicable to formulating-structuring efforts. These include having a clear description of the objectives and goals of the system as a starting point. Since these objectives and goals serve to establish the aims toward which the system is directed, attention should be placed on them initially, with only secondary reference to what limitations practical considerations will place on these goals. Later, the effects of the system environment must be considered in detail to see whether and to what extent it may be necessary to modify the goals initially established.

Each system has a number of inputs and outputs, and it is essential that a good understanding be obtained of all the ones that are significant and of their interrelationships. Not only must the necessary inputs and outputs be specified; deleterious inputs and outputs which should be suppressed must also be stated and their maximum allowable values given.

There are many system structures which are available for effective use by any system. Depending on the equipment, function, organization, time, or other system emphases which are significant, the system structure can take on various forms. A number of examples of representative systems structures have been given to illustrate different practical possibilities that have been shown to be valuable and effective. Ingenuity and commonsense in structuring can be vital assets to most systems work.

4

Factors for Judging the Value of a System

4.0 Introduction

The systems engineered in industry tend to have the characteristics of significance, complexity, and high cost. They are *significant* in that they may contribute in various ways to such things as the productive capacity, the service capability, or the defensive strength of a country and/or company. They are *complex* in that they are composed of many parts which must operate together effectively for the system to perform satisfactorily. They are *expensive* in a cost sense and hence their value must be judged relative to the values of alternative systems or alternative methods of spending the money that would be required for the purchase of the system. In many cases they are one-of-a-kind combinations of several different parts or equipments so that in general it is difficult to determine readily which of various alternative systems is the best choice. Since these systems are expensive they must do what is required of them, for the purchaser is little inclined to spend so much of his money for a system that does not perform satisfactorily.

From this it is apparent that a definite need exists to be able to establish in an organized fashion the value of a system to the purchaser—what are the various possible ways of achieving the desired results and how do these ways compare?[50] Methods of establishing the performance and value of the system are necessary for the purchaser (customer) to judge whether the price (or cost) can be justified. These methods are also necessary for the system maker (vendor) so that the engineer can determine how much cost is likely to be acceptable to the customer, and therefore more specifically what sort of system he should propose and build.

A number of accepted bases for judging the system worth have been described in Chapter 1. They include performance, value, cost, time, reliability, maintenance, and flexibility, that is, the ability to change and grow with the varying environment over a period of time.

This list of factors is merely a representative one and not meant to be considered all inclusive for all applications. Furthermore, the methods and factors available for judging the system will probably change with time. During the initial stages of a system the novelty and conceptual features will frequently figure most dominantly. During the development phases the judgment will be based on limited performance evaluations under simulated conditions; during acceptance of hardware and prototype equipment, the emphasis will be placed on meeting specifications, cost estimates, and delivery times.

A number of these judgment factors, such as cost, time, and reliability, have been touched upon in previous chapters and will be considered in more detail in subsequent ones. The principal emphasis in this chapter will be the value of the system, although performance, flexibility, and maintenance will receive some attention. Two important questions with which this chapter will concern itself are the following: (1) What are the factors which influence one to select a particular set of mission requirements for a system? and (2) What is the value of being able to perform these requirements?

4.1 System Requirements Selected Strongly Influence the Value of a System

Despite the fact that the system has many different kinds of requirements, one or more of these requirements may have higher weight or greater importance than others. The resulting system obtained is often strongly dependent on what is selected initially as being the essential requirements and what relative value is associated with each requirement. As a set of guide lines for establishing these system requirements and determining their relative values, consider the following precepts and questions:

1. The reference or goals for a well-controlled situation establish the general nature of what the results will be.
2. What characteristics would one like the system to have if one were able to have anything he desired?
3. What would one gain if the proposed system turned out just as planned, and what would one lose if the proposed system turned out poorly? What is the likelihood of each possibility?
4. It is necessary to establish the relative importance or weighting of the different system requirements.

5. Since the time needed to design, build, and realize a new system may be many years, the requirements established should be compatible with the conditions which will exist at the time the system will be used.

Since the foregoing are expressed in rather colloquial terms, they warrant some explanation.

Reference Selected Initially Tends to Establish Results

The process of establishing the system requirements serves to single out the factors of importance in the resultant system design. Furthermore, the desired values in some sense, generally a quantitative one, provide goals or objectives toward which the system designers will work. Therefore these desired values of the factors of importance in the resultant design can be likened to the reference in a control system. To the extent that the loop gain of the control is high and the system is stable, the control variable tends to reproduce the reference.

As shown in Figure 4.1-1, the *desired* values of these factors of

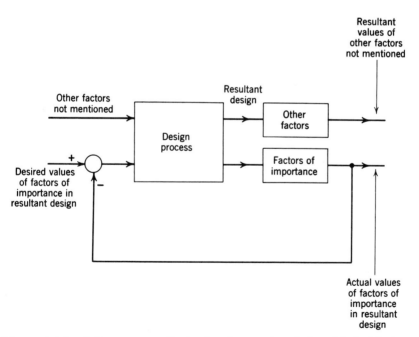

Figure 4.1-1. Influence on resultant valves from system design of desired factors and factors not mentioned.

importance are compared by the system designer to the *actual* values of these same factors in the resultant design. The differences between desired and actual values are used as inputs to the process by which the resultant design is produced. From the resultant design are determined the actual values of the factors of importance of the system to be compared with the corresponding desired values. The design process, as has been pointed out previously in Chapters 1, 2, and 3, is an iterative (closed-loop) one. If the skill and competence of the system designers are reasonably high, i.e., comparable to a high-gain control system, the design process should create a design which will yield the desired values of the factors selected to be of importance.

Referring again to Figure 4.1-1, it will be noted that also serving as inputs to the design process are other factors about the system which are not mentioned specifically as having particular requirements. These factors are also operated on by the same design process and show up in the outputs as resultant values. Since these other factors are not specified, there may be no feedback by the system designer, and the loop tends to operate in an open-loop fashion, that is, whatever value for these factors is produced by the design is accepted as satisfactory and is left unaltered. Frequently at a postmortem of some system design failure, the designer will be heard to protest, "But nobody told me that the system would be required to be operated that way," or "But I was told that I must finish the work within 6 months even if it were not completely satisfactory performancewise."

From the consideration of the system requirements as reference values for the design process, it is apparent that the selection of initial mission and system objectives can have a very powerful influence on the results obtained from any particular system that is designed to meet those objectives.

What Characteristics Would One Like the System to Have?

Having established that the choice for the factors of importance will probably result in a system design that will achieve those objectives, the question now becomes, "What characteristics would one like the system to have if one were able to have anything he desired?" In a sense one is asking, "If I had three or four wishes for the characteristics of this system, what would I choose?" These characteristics represent ideals toward which the designer will direct his efforts.

However, as we will know, in a practical sense one has to say not only what he would like but also how far he is willing to depart from these conditions and in what ways, and still have a set of results that he can consider satisfactory for his purposes.

Ideally one would like a system that has excellent performance in a specific fashion, costs nothing, is presently available, will last forever, requires no maintenance, can be operated by anyone, etc. But in actual fact one is generally quite willing to accept as valuable to him a system which is significantly better than what is presently available or will be available in the foreseeable future. The process may be likened to an optimizing one in which the purchaser selects the best of alternatives available at this time, although this might not be the optimum if he were to wait for values possible far in the future. The following four instances serve to emphasize this point.

1. Although excellent performance is desired, in reality something like 50 or 100 per cent better than what can be done presently may be very acceptable. It should be noted that in some cases the performance requirements are fixed and definite and the relativistic approach implied above is not acceptable. For example, if at the time of the Wright brothers, when flights of 1 or 2 miles were about the limit of the state of the art, a performance requirement of ability to fly the Atlantic Ocean were set forth, an improvement of 100 per cent over the Wright brothers' equipment would not be adequate. Pragmatically speaking, it is interesting to note that the aircraft art did not develop so rapidly with trans-Atlantic flights, but a steady growth through more modest improvements was realized.

2. Although obtaining the desired system for nothing is a pleasant thought, actually for the financial resources which the purchaser involved is able to commit a sum in the range of, e.g., $100,000–$200,000 may be a reasonable and acceptable cost. Such a cost is in general judged relative to costs for other items and may be considered as fair for the service the system performs. In any event, there is generally some cost range within which the customer feels he is able to get a good bargain or return for his investment.

3. Although having the equipment available *right now* sounds like a good idea, the rest of the arrangements, including getting the people required to use the system, can probably not be accomplished immediately. In reality, if the system were ready 6–9 months from now, or at some other appropriate time, this date would serve quite well.

4. Although a desire for the equipment to last forever seems reasonable, on further reflection it is apparent that obsolescence, if only of appearance or of technology, will probably set in after 5–10 years

or at some other foreseeable time in the future. On this basis, an equipment life of 5–10 years may be all that is really needed.

In similar fashion, one can arrive at reasonable maintenance and human factor values that can be accepted as starting points for a study of system requirements and design optimizing and trade-offs.

In some cases judgment values of the sort mentioned above have been well thought out by the system purchaser, who can specify them clearly and with reason. In other cases, it is necessary for the systems engineer to try to establish them initially for himself and then for himself and his customer. In general the process of establishing system objectives should first establish what one would *really* want without considering constraints and then see what effect the presence of constraints has on one's desires.

What Would One Gain Versus What Would One Lose?

The basis of this idea is that if one now knew positively all the future characteristics of the system, and if one knew completely his total future resources and environment, there probably would not be too much of a problem as to what course of action to take. Since these future conditions are not known, one must endeavor to establish the foreseeable results if certain goals or objectives are obtained, as well as the results if they are not obtained.[51] Coupled with this is a risk assessment of the probability of success or failure being realized for each of the objectives.[12]

Figure 4.1-2 shows a decision tree diagram and serves to illustrate the nature of various alternatives associated with the choice of a design method. At 1 one may choose either method A or method B. If one selects method A, he then at 2 can use either new design 3 or old design 7. Having chosen to use the new design, he may do a minimum of preliminary design 4, do a more adequate preliminary design 5, or do a thorough preliminary design 6. For condition 4, the probability of success is lower and the cost is likewise lower. For condition 5, the probability of success is moderate and the cost is moderate. For condition 6, the probability of success is higher and the cost is higher. If the old design, choice 7, is used, the performance and cost are better known and so are the associated probabilities.

If method B is selected, at 8 the choices of whether to buy parts on the outside and do one's own assembly, 9, or to buy assembled units on the outside, 10, must be made. With choice 9, a higher

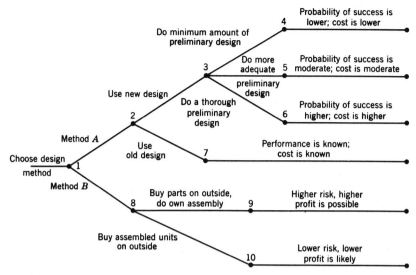

Figure 4.1-2. Decision tree diagram, showing alternative possibilities.

risk is involved and a higher profit appears possible. With choice 10, a lower risk is required but a lower profit is likely.

From a more quantitative consideration of the various alternatives described above, in which numerical values of estimated probabilities of success, costs, profits, risk (also a probability), performance, etc., are employed, the system engineer is able to provide himself with better measures of how successful various alternatives might be as well as a heuristic, or Bayesian, estimate of how likely each alternative is to occur. The degree of sophistication which one cares to employ in considering these alternatives will vary with the importance attached to the decision and the degree of validity of the data associated with the numerical estimates.[24] Basically, to implement this precept, one must be able to *predict*, for example, with the aid of modeling and simulation, what the results will be for the system specified and to establish the *cost* that it will take to obtain these results.

As another example of these considerations, suppose that a plant makes a product valued at $9000 a day with a yield efficiency of 90 per cent. If the efficiency were increased to 95 per cent by the addition of some new equipment, there would be an increased income of $500/day. Over a period of a 2-year life for the new equipment, this would result in an increased income or gain of

$$365 \times 2 \times \$500 = \$365,000$$

If, however, the new equipment has a reliability of 80 per cent and its failure to operate would shut down the original equipment, then the daily income will be 9500 × 0.80 = $7600/day. Hence the addition of new equipment of this low reliability would result in a decreased daily income of 9000 − 7600 = $1400/day.

The possible gain of $500/day would thus fail to be realized, and a loss would be incurred instead. Obviously a much higher reliability requirement or a design which would not interfere with the operation of the original system if the new equipment failed is needed.

Even with a reliability figure of 100 per cent, one would still have to compare the gain of $365,000 for 2 years with the total system cost. If this system cost were $100,000, the answer would probably be favorable; if the cost were $500,000, the answer would be likely to be negative.

Although these figures have been purposefully chosen to be either black or white, they serve to illustrate the process of comparing the alternative values to be achieved for various objectives and balancing the gain or loss for the alternative methods. In effect, this process is one of best-case and worst-case estimating and can prove very effective in crystallizing one's opinion of alternative methods.

Establish Importance of Different System Requirements

The preceding precept of balancing gains and losses attributable to different system requirements indicates the need for having some way of equating the relative worth of these quantities with different sets of units. Traditionally, we have been told that we cannot compare unlike things, and it would appear that system factors such as performance, time, and reliability have quite dissimilar units. Fortunately, if each of them can be related to a common term, for example, the overall worth or output of the system, it is then possible to establish their relative worth.

Referring to Figure 4.1-3, if there are two different inputs to the system, each of which contributes to its overall utility, it is possible to express the utility as a function of the two variables:

$$U = f(A, B) \tag{4.1-1}$$

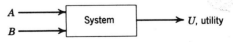

Figure 4.1-3. System utility shown as a function of different inputs, A and B.

If the partial derivatives are taken,

$$dU = \frac{\partial f(A, B)}{\partial A} dA + \frac{\partial f(A, B)}{\partial B} dB \qquad (4.1\text{-}2)$$

For comparable utility to exist, dU for dA and dU for dB must be equal so that

$$dA = \frac{\dfrac{\partial f(A, B)}{\partial B}}{\dfrac{\partial f(A, B)}{\partial A}} dB \qquad (4.1\text{-}3)$$

where

$$\frac{\partial f(A, B)/\partial B}{\partial f(A, B)/\partial A} = W_{B/A} = \text{relative weighting of } B \text{ to } A$$

In the general case, comparable to Equation 4.1-1, the utility U is a function of many variables, and the incremental utility ΔU from Equation 4.1-2 can be written as

$$\Delta U = \sum_{n=1}^{n=N} W_n \Delta Q_n \qquad (4.1\text{-}4)$$

where ΔU is the total incremental utility,

ΔQ_n is the change in each of the input variable quantities,

and

W_n is the weighting factor corresponding to each of the input quantities.

Although it is desirable that the weighting factors be absolute, frequently one is required to settle for relative weighting of one quantity to the rest. Frequently, this does not really pose any significant problem, and relative values are good enough.

A common basis of utility must be arrived at to permit the weighting factor evaluation. This utility base is not always the same for each system or for all systems that might be considered. Money or performance or even a ratio of the two, performance/$, is frequently used as a common basis for expressing utility.

Include Effect of Changes Likely to Take Place during System Development

Since the span of time involved during which a system is developed and brought into being may be long, i.e., measured in years, and

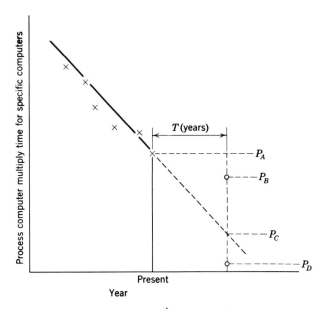

Figure 4.1-4. Process computer multiply time versus years.

since the rate of change of technological development during those years may be rapid, it is important that the system objectives or requirements be compatible with the conditions which will exist when the system appears or is operative.[54] It is not sufficient that they be satisfactory for present conditions; they must also be suitable for the time, T years from now, when the system is to be used.

As an illustration of this dynamic effect, consider Figure 4.1-4, which shows a plot of process computer multiply times for specific computers as a function of years for the several years of their early development up to the present. If at the present, when a performance P_A is obtainable, one establishes a performance objective of P_B, one might think that a desirable objective had been set. However, if the present improvement rate in process computer multiply times is maintained (and the likelihood of such continued improvement should be ascertained) during the next T years when the objective P_B is being sought and realized, the performance level required to be competitive is likely to be P_C. To be superior in performance at that time, some such performance objective as P_D may be required. Although it may not be essential for P_D to be realized, the likelihood of P_B being a competitive value of performance at the time when the new process computer system is available is not high.

Performance as a Measure of Value 145

In leaving the subject of establishing the system requirements, let us again remind ourselves that this part of the job is extremely important. It provides the standard on which all subsequent effort is based.

4.2 Performance as a Measure of Value

Once the requirements for the overall system and for the corresponding subsystems have been established, it is essential that every effort be directed at meeting these requirements. One of the most important is performance because, in the final analysis, the performance represents the reason why the system was built in the first place. Hence, as a rough approximation the value of the system is strongly a function of its performance.

Thus, as in a race or other fair competition, the individual or system with the best performance is declared the winner and is entitled to greater recognition and reward than the runner-up. Other factors, such as cost, time, and reliability, in a way tend to represent constraints which must be met without prejudicing performance too severely, and these factors will be discussed in subsequent chapters.

Generally a certain level of performance is needed to justify the worth of the system, even if this serves merely to demonstrate the feasibility of performing the desired task. However, in many cases, merely being able to meet the performance needs is not enough because all of the companies or organizations offering to make the system claim this ability. Under circumstances where the customer has specified the performance level required, the other factors of cost, time, etc., may no longer serve only as constraints. These factors may then represent principal means of comparison for deciding which of the rival organizations should be selected to do the systems job.

Because determining the performance of the components of the system as well as of the system itself is strongly dependent on the particular type of system being considered, and because for each type of system the subject matter required to describe the performance may fill many volumes of college-level texts and beyond, no effort will be made here to cover in any detail the methods of establishing the calculations of performance. However, several examples of typical ways of describing performance will be presented briefly so that an idea may be obtained of some of the widely varied ways in which performance criteria have been set forth.

Figures of Merit

A highly desirable way of expressing performance is in terms of non-dimensional form in which the size or number of variables does not completely mask the significance or generality of the results. Frequently such terms, which tend to compensate for size or number of units involved, are referred to as figures of merit, e.g., the torque to inertia ratio, T/J, or the damping ratio, ζ.[15] Performance indices, such as these figures of merit, provide a way of relating the performance of systems in different fields or different sizes on a somewhat comparable basis. A number of other nondimensional terms or ratios used in several different branches of engineering are described below.

The use of ratios and other figures of merit can help in avoiding the situation of the farmer who was concerned that his white horses were eating more than his black horses until someone pointed out he had many more white than black horses.

Electrical Performance

Efficiency.[49] A common basis for describing electrical performance in the energy field is efficiency. By expressing both the efficiency and load in percentages, curves such as those for A and B type motors in Figure 4.2-1 are obtained. Curves of particular shapes are characteristic of certain types of equipment performance, and it is possible

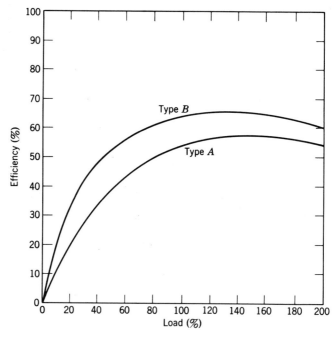

Figure 4.2-1. Comparison in per cent of efficiency of two motor types versus load.

to estimate the loss requirements for an unknown equipment on the basis of a knowledge of the approximate shape of the efficiency curves for the appropriate class of equipment. Furthermore, by noting the shape of the efficiency curve over the entire range of loads, it is possible to evaluate the total amount of loss for a particular system duty cycle.

Per Unit. Another convenient method for describing electrical equipment performance is the use of the per unit system.[1] The basic electrical equations

$$E = IR \tag{4.2-1}$$

and

$$P = EI = I^2R \tag{4.2-2}$$

may be referred to a common voltage and current base and expressed as a ratio to these base quantities.

Thus, if

$$E_b = \text{base voltage} \quad \text{and} \quad I_b = \text{base current} \tag{4.2-3}$$

then

$$e_{\text{in per unit}} = \frac{E}{E_b} \quad \text{and} \quad i_{\text{in per unit}} = \frac{I}{I_b} \tag{4.2-4}$$

With the values of E_b and I_b chosen, the resistance and power bases are then fixed. Thus

$$R_b = \frac{E_b}{I_b} \quad \text{and} \quad P_b = E_b I_b \tag{4.2-5}$$

Not only resistances but reactances as well can be expressed in per unit.

By means of the per unit system, one is able to perform electrical network analysis using simple numbers about unity for many of the calculations required. A further advantage of the per unit system is that similar machines and equipments of different sizes have been shown to have a similar range of per unit reactances. Table 4.2-1

Table 4.2-1 *Values of Reactance (In Per Unit)*

Size of Alternator (kw)	Synchronous	Transient	Subtransient
100–2500	1.50–3.00	0.15–0.45	0.09–0.27
2500–150,000	1.41	0.13	0.10
150,000–500,000	1.65	0.25	0.18

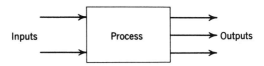

Figure 4.2-2. Generalized chemical process input-output diagram.

shows the values of synchronous, transient, and subtransient reactances of electrical alternators for a broad range of different sizes of machines.

Chemical Performance

In many chemical engineering processes one or more input quantities are supplied to the process and two or more output quantities are obtained; see Figure 4.2-2. The performance of these processes can be judged on a number of different bases, but two important ones are conversion and yield. *Conversion* is the term used to relate the amount of a given material going out of the process to the amount coming in.

$$\text{Conversion} = \frac{\text{Amount of material going out}}{\text{Amount of material coming in}} \quad (4.2\text{-}6)$$

Figure 4.2-3 shows a plot of conversion versus reaction temperatures.

Yield is the term used to relate the amount of desired product to the total output. This is determined on the basis of the amount of product produced per unit of time. Thus

$$\text{Yield}_{\text{per unit time}} = \frac{\text{Amount of desired product}}{\text{Amount of total output}} \quad (4.2\text{-}7)$$

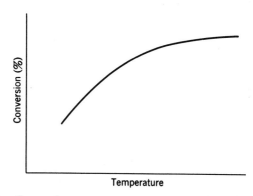

Figure 4.2-3. Conversion versus temperature for a given chemical process.

Performance as a Measure of Value

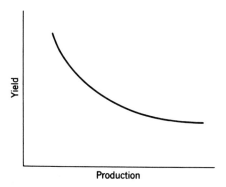

Figure 4.2-4. Yield versus production for a given chemical process.

Figure 4.2-4 shows a plot of yield versus production and shows that as more of the desired product is produced this is done at the expense of a less efficient production rate.

Thermal Performance

Over the years the size of steam-turbine generator units has increased very markedly from early ones of 5000 kw to present-day ones of 1,000,000 kw. Likewise the range of steam temperature over which the turbines have operated has changed appreciably. It is

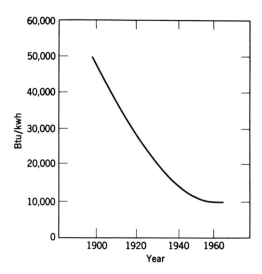

Figure 4.2-5. Btu/kwh for most efficient power plant as a function of time in years.

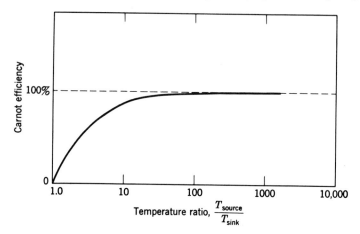

Figure 4.2-6. Carnot thermal efficiency as a function of source/sink temperature ratio.

desirable to have ways of comparing the performance of present-day units with those of previous times. In this fashion it is possible to develop performance versus time in years characteristics by means of which future performance objectives may be more intelligently anticipated.

Figure 4.2-5 shows a plot of the heat energy required to generate a kilowatt-hour in the most efficient plant over a number of years. Thus, despite the many-orders-of-magnitude change in power range of the units involved, the amount of energy required to generate a kilowatt-hour has been gradually improving as a result of continued technical progress.

A very useful concept in thermal energy consideration is the Carnot efficiency of a heat cycle. Carnot showed long ago that the efficiency of thermal conversion is limited by the ratio of the highest temperature to the lowest temeprature. Figure 4.2-6 shows a plot of Carnot efficiency as a function of the temperature ratio.

Material Performance

The performance of materials tends to be on more of an absolute than a relative basis, although frequently it is expressed as a ratio to nominal or rated values. Figures 4.2-7 to 4.2-9 show some representative performance curves for various typical characteristics of three different materials.

Figure 4.2-7 shows how the resistance ratio of two cryogenic mate-

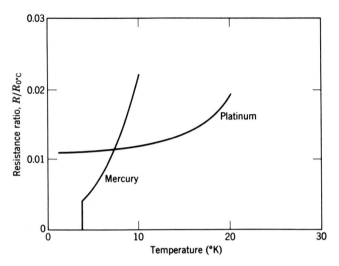

Figure 4.2-7. Resistance ratio for certain metals as a function of temperature.

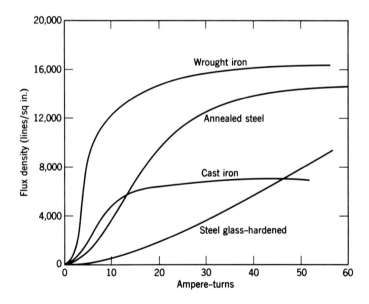

Figure 4.2-8. Flux density versus ampere-turns for several different grades of ferromagnetic materials.

151

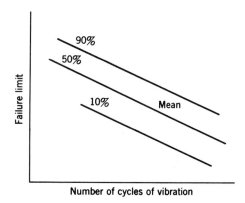

Figure 4.2-9. Failure limit for 10, 50, and 90 per cent conditions of material as a function of number of cycles of vibration.

rials, compared to their resistances at 0°C, changes as a function of temperature in the vicinity of absolute zero.

Figure 4.2-8 shows that the flux density (lines/square inch) of certain common magnetic materials varies as a function of the number of ampere-turns of excitation applied. The use of ampere-turns is somewhat nondimensionalized, since the abscissa permits one to handle equipment of different current ratings and physical sizes with the same charts.

Another field of materials interest is illustrated by Figure 4.2-9, where the failure limit for a material or member is shown as a function of the number of cycles of vibration. Curves are drawn for the 10, 50, and 90 per cent limits.

Transportation Performance

In addition to performance indices or figures of merit for equipment built around a particular technical discipline, there are comparable ways of indicating performance and/or efficiency for functional tasks, e.g., transportation. Von Karman and Gabrielli[2] have provided data which compare alternative means, such as submarines, ships, automobiles, aircraft, and autorail, from the point of view of transport efficiency versus maximum speed in miles per hour, where

$$\text{Transport efficiency} = \frac{WV}{P} \qquad (4.2\text{-}8)$$

with W = gross weight of craft,
 V = design of speed of craft,
and P = total installed horsepower.

Performance as a Measure of Value

Figure 4.2-10 shows a plot of the data from Von Karman and Gabrielli and provides a basis for comparing the performance of these different methods. Future improvements may of course modify such curves, but if a consistent time base is used, the relative transport efficiency as so defined of new means such as hydrofoils or ground-effect machines (GEM's) may be compared with that of prior equipments.

Control Performance

Because the various portions of control systems have involved the use of many different forms of energy at many different power levels,

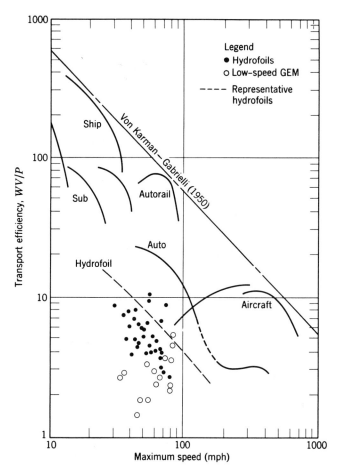

Figure 4.2-10. Transport efficiency versus maximum speed for different means of transportation.

typical representations of control performance frequently are in the form of dimensionless ratios. Although frequencies are in many cases plotted to logarithmic scales, time is generally expressed on a fixed base. Figure 4.2-11 shows the open-loop and closed-loop frequency responses for a general position control as well as the corresponding

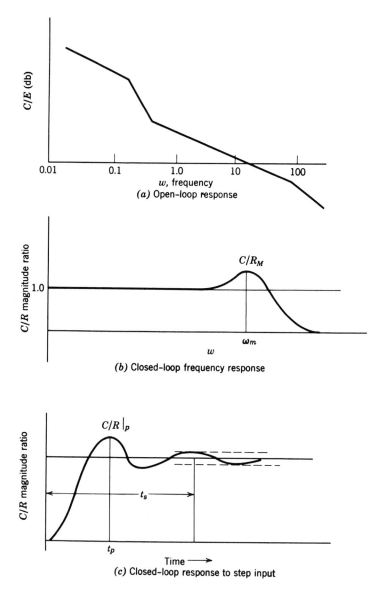

Figure 4.2-11. Nondimensionalized frequency response and transient response means for expressing dynamic control performance.

response as a function of time to a step input. Charts showing nondimensional ratios for relating transient and steady-state responses have been developed and can be used to advantage.[15]

From this brief survey of typical performance representations for various kinds of systems, it is apparent that systematic ways of generalizing performance results into nondimensional ratios, per unit systems, and their figures of merit are useful in obtaining a broader perspective on the performance of representative classes of equipments and materials. Their use can be very helpful to the designer in establishing what his systems should be required to do or what other people's system are likely to do now and in the future.

4.3 Net Return Basis for System Evaluation

Apart from their performance capabilities most systems are also evaluated from a financial viewpoint.[32] One practical and pragmatic way of judging the value or worth of a system is the net return which is obtained; namely, how much money the system returns for the costs involved.[60] Depending on the particular type of system and the magnitude of the risks and resources involved, the net return which will constitute a satisfactory value will vary. However, the net return is a financial criterion which is well understood by the type of management people who have to approve the venture before formal work on the system can be undertaken and it is desirable to be able to express the net return for a system when this is appropriate.

Stated in a simple form,

$$\text{Total net return} = \text{Total value of product or service produced} \\ - \text{Total cost of product or service} \quad (4.3\text{-}1)$$

where the value of the product or service produced indicates what the purchaser is willing to pay for the system, and the cost indicates what the producer is required to pay in order to make the system and bring it into usefulness.

At first glance it might seem that this simple difference of two terms involves a straightforward mathematical operation and that the determination of the total net return is a clear-cut basis for obtaining a unique figure for the worth of a system. However, a number of factors involved in both the value and the cost of the product are functions both of time and of the resources of the producer and the purchaser of the system.

It is found that the most effective way of evaluating Equation 4.3-1 is to subdivide the total value and total cost terms into a consistent set of factors which go to make up each of these totals. Because the value and the cost are comprised of many terms and because value and cost are each a fairly comprehensive topic by itself, a further discussion of the value of a system will be presented later in this chapter, while a more detailed treatment of the cost of the system will be given in Chapter 5. Although the value and the cost of a system are obviously related and must be considered jointly in a systems study, from the point of view of simplicity and clarity it is worthwhile to describe them somewhat separately.

Time History of System Costs and Value Produced

To illustrate the nature of the total costs, the value of product produced, and therefore the total return as a function of time for a typical system,[53] Figure 4.3-1 shows the dollar return from a system, as well as its value and cost terms, from the start of the project. At

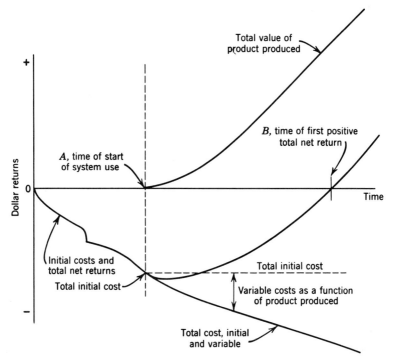

Figure 4.3-1. System costs and value as a function of time from initiation of project.

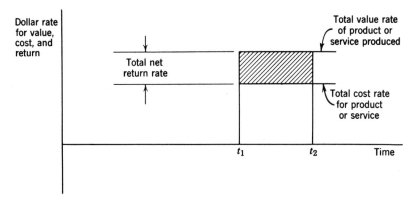

Figure 4.3-2. System value and cost rates as a function of time during steady-state conditions.

the outset, there are only initial costs and a negative total net return of an increasing amount until the time of start of system use, A. By this time, the total initial cost may have been expended. As initial productive use of the system begins, the value of the product produced becomes positive, probably at a low rate at first, then at an increasing rate as the system begins to realize its design capabilities. Although the total initial cost at the start of system production is shown to remain unchanged, there are additional variable costs as a function of the product produced. The total net return, as given by Equation 4.3-1, remains negative for some time after the start of system use because of the large initial cost as well as the added variable costs. The total net return is shown to become positive at B, designated as the "time of first positive total net return."

Presumably after time B the total net return will continue to increase, and the economic value for the system becomes positive and significant. The transient net return versus time phenomena shown on Figure 4.3-1 is part of what is referred to as cash flow by businessmen and emphasizes the importance of cost and time in judging a system. Since for some systems the time from the start of the project to the first positive total net return may be measured in years, and since the magnitude of the maximum negative total net return may be significant, the total net return considerations provide a relatively important means with which to evaluate a system.[53]

In its steady state, a system may have essentially a constant total rate of value of product or service produced, of cost for product or service, and of net return, as shown by Figure 4.3-2. For this ideal condition the total net return rate times the time interval, $t_2 - t_1$, i.e., the shaded area, would yield the total net return for that period. However, to evaluate the overall total net return from the system as given by Equation 4.3-1, one needs knowledge or reasonably good estimates of the answers to many questions, such as the following.

Over how long a period of time should the total system effort be considered to extend? What will be the effectiveness of the system over this period? What short-time expenditures or costs can be incurred without jeopardizing the possibilities of long-time gains?

What is the total value of the product or service produced? What price or prices can be charged for the product or service, and what are the number of units or people which can be handled by the system? In general, what is the proper size for this system? What should be the size for this system relative to past and future systems? Obviously, the total value of the product or service produced places an upper limit on the total net return.

Before the total cost of the product or service can be established, it is essential that the size of the system be determined. However, certain costs and expenditures will be incurred somewhat independently of how large, within fairly broad limits, the system is. Obviously, the total cost of the product or service places a lower limit on the total value below which the total net return is negative. An iterative trial-and-error procedure is required to establish the total value and the total cost on a comparable basis to make it possible for a positive net return to be indicated. Although no *correct* answer is likely to be obtained, reasonable estimates can be determined as a basis for judgment of the potential return of the system.

Present-Worth Concept

The preceding description of the time history of system costs and value produced has served to show the dynamic nature of the total net return as a function of time. Another factor of significance in considering total net return as a function of time is the changing value of money with time, arising from the possibility of obtaining interest by depositing the money in a bank or placing it in an alternate investment as described by Grant and Ireson.[23] Thus, under the basic system concept that there are alternative methods for accomplishing a given result, the person or organization investing his money in a system has the alternative of placing his money in another venture whereby he can increase its value by a known amount as a function of time.

For example, by placing P, a present sum of dollars now, in a

Net Return Basis for System Evaluation

bank having an interest rate of i per cent per period, after n periods the investor will receive a future sum of dollars, S, given by the formula

$$S = P(1 + i)^n \qquad (4.3\text{-}2)$$

or

$$P = \frac{S}{(1 + i)^n} \qquad (4.3\text{-}3)$$

The upper family of curves on Figure 4.3-3 shows how the original amount of money, P, grows over the years by means of compound interest so that S, the future worth of P now, is greater than P depending on the interest rate, i. Table 4.3-1 shows the single payment compound-amount factor, the ratio of S/P of Equation 4.3-2, as a function of the number of periods, n, during which the present sum P is held at interest.

The converse of this idea, that the present sum, P, is equal to the sum S, n periods from now, is that the sum S' in the future is worth the present sum P', now. On Figure 4.3-3, for example, if

Table 4.3-1. *Single Payment Compound-Amount Factor*

$$\frac{S}{P} = (1 + i)^n$$

for Various Values of Interest Rate, i, and Number of Periods, n

Number of Periods, n	Interest Rate, i (%)						
	2	3	4	5	6	8	10
1	1.020	1.030	1.040	1.050	1.060	1.080	1.100
2	1.040	1.061	1.082	1.103	1.124	1.166	1.210
3	1.061	1.093	1.125	1.158	1.191	1.260	1.331
4	1.082	1.126	1.170	1.216	1.262	1.360	1.464
5	1.104	1.159	1.217	1.276	1.338	1.469	1.611
6	1.126	1.194	1.265	1.340	1.419	1.587	1.772
8	1.172	1.267	1.369	1.477	1.594	1.851	2.144
10	1.219	1.344	1.480	1.629	1.791	2.159	2.594
12	1.268	1.426	1.601	1.796	2.012	2.518	3.138
15	1.346	1.558	1.801	2.079	2.397	3.172	4.177
20	1.486	1.806	2.191	2.653	3.207	4.661	6.727

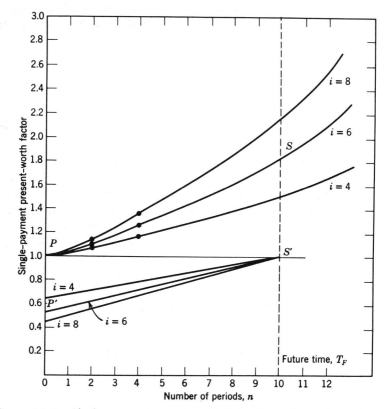

Figure 4.3-3. Single-payment present-worth factor, for various values of interest rate, i, and number of periods, n.

the future value of the sum after 10 years is chosen to be $S' = P$, the lower family of curves shows that the present sum P', now, will be less, depending in amount on the interest rate, i. Table 4.3-2 shows the single payment present-worth factor, the ratio of S'/P' as a function of the periods in the future, n, and the interest rate, i, for Equation 4.3-3.

It is this effect of compound interest and time which gives rise to the *present-worth concept* illustrated above. This fundamental fact that the value of money is a function of the time when the money is required is an influencing factor in considering the values of alternative systems. Thus, the value of the total net return is a function of the time at which one realizes it. More lengthy discussions of the present-worth concept are to be found in such books

Table 4.3-2. *Single Payment Present-Worth Factor*

$$\frac{P}{S} = \frac{1}{(1+i)^n}$$

for Various Values of Interest Rate, i, and Number of Periods, n

Number of Periods, n	Interest Rate, i						
	2	3	4	5	6	8	10
1	0.9804	0.9709	0.9615	0.9524	0.9434	0.9259	0.9091
2	0.9612	0.9426	0.9246	0.9070	0.8900	0.8573	0.8264
3	0.9423	0.9151	0.8890	0.8638	0.8396	0.7938	0.7513
4	0.9238	0.8885	0.8548	0.8227	0.7921	0.7350	0.6830
5	0.9057	0.8626	0.8219	0.7835	0.7473	0.6806	0.6209
6	0.8880	0.8375	0.7903	0.7462	0.7050	0.6302	0.5645
8	0.8535	0.7894	0.7307	0.6768	0.6274	0.5403	0.4665
10	0.8203	0.7441	0.6756	0.6139	0.5584	0.4632	0.3855
12	0.7885	0.7014	0.6246	0.5568	0.4970	0.3971	0.3186
15	0.7430	0.6419	0.5553	0.4810	0.4173	0.3152	0.2394
20	0.6730	0.5537	0.4564	0.3769	0.3118	0.2145	0.1486

* From Grant and Ireson *Principles of Engineering Economy*.[23]

as Grant and Ireson[23] and other standard textbooks on engineering economics.

4.4 Value of a System

One of our well-established bases for setting the value of a system is the classical economic concept which says in effect, "The value of something is what someone is willing to pay for it." More precisely, "Value is a fair return in money, goods, or services for something exchanged." Because there exists a disparity, or interface, between what the goods or services mean to the individual or group supplying them and what they mean to the individual or group purchasing them, a condition can exist where both the purchaser and the supplier are able to benefit for an exchange at a particular price.

Such situations exist because the skills and abilities of the system supplier are able to give the purchaser the capability to accomplish things that in turn have a value, perhaps in another frame of reference, that is in excess of what the purchaser has to pay. For example, the skill required to build an electronic computer capable of handling airline reservations may be well within the routine capabilities of those skilled in this art; however, the ready inclusion of such a reservation system in an airline system may provide far more rapid, accurate, and reliable service than could be obtained previously with a manual-telephone system. On the other hand, having such a computer hardly would enable the supplier of the machine to go into the airline business.

Three different ways of looking at the value of a system will be considered at this time: (1) the demand-supply viewpoint of classical economics, (2) the individual's demand-income point of view based upon empirical studies of the market, and (3) the value-added concept of the maker. Each of these methods may provide a different answer as to what the value of a particular good or service should be, and it is impossible to say with assurance that one is more valid than another. Since the systems engineered by industry may be sold under a variety of conditions, some one of these methods may be more advantageous for use than the others for a particular situation. It is therefore worthwhile to be familiar with all three methods.

Demand-Supply

In the classical economics concept of the market at which a particular product is being sold there are a large number of buyers and any amount of the product that is desired can be made. Furthermore, each unit of the product has essentially the same qualities as the next. Presumably, the buyer who buys the first unit is the one who is willing to pay the most for it. Succeeding buyers may be willing to pay for later units as much as the first buyer paid or less.

From these assumptions, one is able to plot (Figure 4.4-1) the total demand expressed in sales as a function of the supply in units for several different supply-demand relationships. A number of different conditions may exist for such markets. Curve A shows a situation where the demand is proportional to the supply and the value of each additional unit is essentially the same as that of the previous one. Curves B, C, and D show in increasing order of severity conditions where, as the supply increases, the demand falls off and the value of each unit is somewhat less than that of the previous one. The degree to which the incremental demand falls off as the supply is increased is referred to as the *elasticity* of the demand.

Figure 4.4-2 shows a plot of the slope of the total demand versus supply curves of the previous figure, which indicates the incremental

Value of a System 163

demand as a function of supply. For a product with a *highly elastic* demand, curve D, the demand for each succeeding unit is considerably less than for the previous one. This is in contrast with curve A, which is the condition of a highly *inelastic* market; here the demand or value for each succeeding unit is essentially the same as for the previous one. In between are varying degrees of elasticity of demand, curves B and C.

Although there is no sure way of reliably establishing a priori the nature of the incremental demand-supply curve, such as that shown on Figure 4.4-2 for a new type of system with a limited number of potential buyers, the person pricing a system engineered in industry is faced with the need of estimating the equivalent of such a curve with only approximate and generally uncertain data. Intuitive estimates handled in a Bayesian fashion[62] can be employed to generate such supply-demand curves and their corresponding incremental demand curve counterparts.

These curves provide a basis on which to estimate the return that may be realized from the sale of increasing numbers of systems. From this information, an estimate of the cost of manufacture, and an idea of what return is required on the investment, it is possible to obtain an idea of how many systems should be built and to place a price on the value of each system.

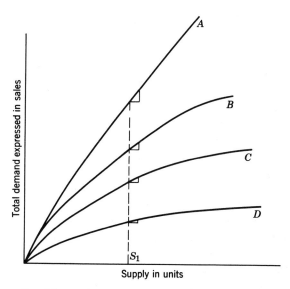

Figure 4.4-1. Total demand as a function of supply for different types of market elasticity.

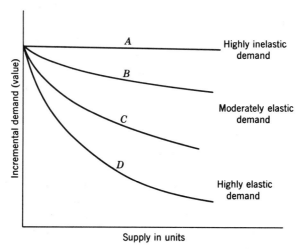

Figure 4.4-2. Incremental demand as a function of supply for different types of demand elasticity.

Demand-Income

From a practical viewpoint, it is evident that an individual (and/or a company) finds his demands or values markedly affected by the amount of income or money he has available to spend. The size of income then may be an approximate indication of how much money the individual can spend and therefore of his demand for a product or service.

Hall has described this condition well in his book, *A Methodology for Systems Engineering*.[34] He indicates intuitively that, as a consumer's income rises, he will, in reallocating his expenditures, spend relatively more on goods which he regards as luxuries than on those which he considers necessities.

The term luxuries and necessities are employed in the usual sense, but they also have a particular economic significance as they relate to the quality of elasticity. A *necessity* is a product which is needed regardless of price, that is, a person will buy it with his first element of income, and therefore the demand with income for a necessity is inelastic. A *luxury* is a product which is desirable but not absolutely necessary. A person may buy a luxury if he has the money and thinks the price is favorable. Hence luxuries tend to have an elastic demand with income.

A third term, an *inferior good*, is also important in this regard and refers to an item for which another good is substituted as income

Value of a System

rises. The term inferior good does not reflect on whether or not the product will do the job required of it, but rather indicates that less, not more, of it will be sold as a person's income rises. Associated with inferior goods is a negative income elasticity, i.e., decreased purchases with higher income. Examples of such goods in a family food budget might be oleomargarine and condensed milk, which would be replaced by butter and whole milk if the latter could be afforded.

Tornqvist[4] has made extensive studies of family budget data in Finland and Sweden. He found four types of curves that have the great merit of being applicable over wide ranges of income. They are:

Necessities: $\quad q_1 = a_1 \dfrac{\mu}{\mu + b_1}, \quad a_1, b_1 > 0$

Relative luxuries: $q_2 = a_2 \dfrac{\mu - c_2}{\mu + b_2}, \quad a_2, c_2 > 0, \; b_2 > -c_2$

Luxuries: $\quad q_3 = a_3 \mu \dfrac{\mu - c_3}{\mu + b_3}, \quad a_3, c_3 > 0, \; b_3 > -c_3 \quad (4.4\text{-}1)$

Inferior goods: $q_4 = a_4 \dfrac{\mu - c_4}{\mu - b_4}, \quad a_4 > 0, \quad \mu > b_4 > c_4$

where the q's are the quantity of the demand, and μ indicates the income. Figure 4.4-3 shows the general forms of these equations. The dotted lines are asymptotes to which the function tends at high

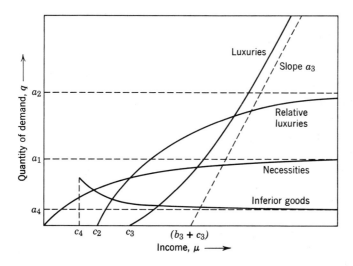

Figure 4.4-3. Tornqvist's demand functions.

levels of income. The curves satisfy our intuition about the way most individuals behave at various income levels. When incomes are low, most expenditures are made for food, basic shelter, and minimal clothing. As income increases, the proportion spent on these items decreases. Above a certain income, certain relative luxuries are afforded. For very high incomes the outlay for luxury items such as yachts, boarding schools, and expensive cars tends to be proportional to income.

From the point of view of determining the value of a system, there are several interesting implications to this demand-income concept. For example, since the terms luxuries, relative luxuries, necessitities, and inferior goods are relative, the customer's actual demand may be strongly dependent on whether he can be made to think that a particular purchase is a necessity or a relative luxury. Furthermore, if in his mind the product is a relative luxury, whether it is priced above or below c_2 may make an appreciable difference in his actual demand. Hence, by correctly estimating the customer's c_2, the seller can establish his prices accordingly. Another interesting point is that beyond a certain income or financial position the customer may no longer be interested in purchasing inferior goods. He may be willing to pay more for a "necessity" which, though more expensive, performs a better job.

Although the demand-income method of establishing value is a rather subjective one, it serves to bring out certain important system-pricing principles that exist even though they are difficult to quantify precisely.

Value-Added

The value-added concept is built on the basis that value is added to a material through processing. Figure 4.4-4 is an illustration of the format of the value-added classification. In this concept, all products are viewed as being made of materials that are orginally provided by Nature—natural resource materials in the conditions and places in which Nature provides them. Only rarely are such natural materials immediately marketable products in the usual sense of the term. As a rule, something must be done to them to make them into products for which there is a demand. This may require only their transportation to the market. Generally, however, such materials require the application of one or more processes to convert them into other materials or products of higher order that customers want. It is the intent of Figure 4.4-4 to suggest that suitable processing confers added value, and that the result at any stage may be sold as

Value of a System

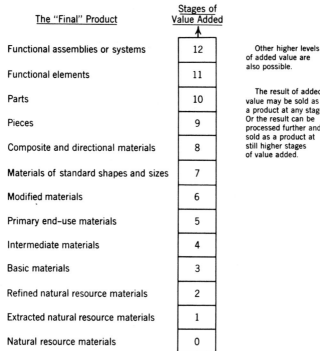

Figure 4.4-4. The value-added classification scale.

a product or may be processed further and sold at some still higher stage of value-added. Dudley Chambers has developed some useful illustrations of this value-added concept which are described below.

Because systems themselves fit into this hierarchy at the higher echelons of added value and also because systems are frequently required to operate with processes and/or equipment at the lower stages of added value, it is worthwhile to define more specifically what occurs at the various levels of value-added. Table 4.4-1 shows examples of the value-added classification of products and services, and brief definitions of the different stages are given below.

Natural resource material (0): a material in the condition and place in which nature provides it.

Extracted natural resource material (1): a material that has been removed from its source so that it can be processed further. The mining of coal and the cutting of logs are representative of the extraction processes.

Table 4.4-1. *Example of Value-Added Classification of Products and Services*

Examples by Type of Customers

Stage of Value Added	Basic Classification	Defense	Industrial	Utility	Original Equipment Manufacturers, and Contractors	Consumers
19	8th order complex functional assembly	U.S. Air Force				
18	7th order complex functional assembly	Strategic Air Command	The several works, mines, and other facilities of an integrated steel company	The system formed by interconnecting the systems of electric utility companies		
17	6th order complex functional assembly	Operating wing—air to ground offense	A steel works for producing various steel products	An electric utility's system		
16	5th order complex functional assembly	Bomber	A hot strip mill or plant	An electric utility's generating station	Milling machine	House
15	4th order complex functional assembly	Bombing system for bomber	Electric power system for hot strip mill	Electric system for utility station	Electrical equipment for milling machine	Electric system for house
14	3rd order complex functional assembly	Radar equipment	Rectifier and switchgear assembly	Unit substation	Speed variator	Packaged electrical kitchen
13	2nd order complex functional assembly	Radar transmitter	Rectifier equipment	Switchgear	Control panel	Refrigerator
12	1st order complex functional assembly	Modulator	Cooling system for rectifier	Circuit breaker assembly	Regulator	Refrigerating machine
11	Functional element	Pulse transformer	Motor	Current transformer	Contactor	Motor
10	Part	Core assembly	Stator assembly		Armature	Stator assembly
9	Piece	Formed coil	Lamination		Sintered contact	Formed coil
8	Composite and directional material	Insulated wire	Oriented sheet steel (if used)			Insulated wire
7	Material in standard shape and size	Copper rod and wire	Sheet steel and strip			Copper rod and wire
6	Modified material			Butyl rubber compound (includes fillers)		
5	Primary end-use material		Steel ingot	Copolymerized butyl gum	Silver powder and tungsten carbide powder	
4	Intermediate material			Butadiene and isobutylene		
3	Basic material	Copper ingot, billet, or bar	Pig iron	Butane and isobutane	Silver ingot	Copper, ingot, billet, or bar
2	Refined natural resource material	Refined ore	Refined ore	Petroleum distillates	Electrolytic silver	Refined ore
1	Extracted natural resource material	Extracted ore	Extracted ore	Crude oil	Crude silver	Extracted ore
0	Natural resource material	Copper ore in the ground	Iron ore in the ground	Oil in the ground	Silver-bearing ore in the ground	Copper ore in the ground

Refined natural resource material (2): a material that has been processed into separately identifiable substances. Ores separated into the desired minerals or mineral oils prepared from distillates are examples of refined materials.

Basic material (3): a material manufactured by separating a desired material from its refined natural material. Iron produced from iron-ore concentrates and chlorine from salt are representative basic materials.

Intermediate material (4): a material obtained from combining basic materials to produce particular qualities that have broad utility in subsequent conversion operations. Dye intermediates produced from benzene are an example.

Primary end-use material (5): a material characterized by specific physical properties in addition to its chemical composition.

Modified material (6): a material combining the special properties of several other materials. Rubber compounds produced by combining rubber polymers and reinforcing fillers represent such modified materials.

Material in standard shapes and sizes (7): a material processed into forms that can be easily and conveniently manipulated for producing further products.

Composite and directional material (8): a material having physical properties purposefully not homogeneous throughout its mass. Enamelled wire and tin plate are composite materials; oriented magnetic sheet iron is a directional material.

Piece (9): a separate and distinct entity of material that has been subjected to processes that have added a specific shape, dimensions, and properties. Pieces may be used as components or as connections.

Part (10): an assembly of two or more pieces that have been joined more or less permanently. Electric fuses, fixed resistors, and plastic laminates are all parts.

Functional element (11): an assembly of parts, pieces, and processed material that will perform a specified function which cannot be performed by any of those constituents alone.

Functional assembly or system (12-19): a product or product constitutent that will perform functions which cannot be performed by any of its constituents alone, created by properly assembling or connecting two or more suitable functional elements and other suitable constituents. The act of creating by "properly assembling" of suitable functional elements and constituents represents a part of the work required and the value added by systems engineering.

Figure 4.4-5 shows in a schematic fashion how component pieces,

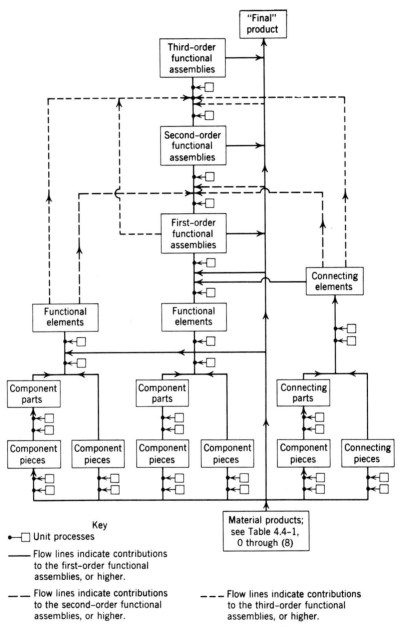

Figure 4.4-5. Products and their constituents.

parts, functional elements, and connecting elements are put together to form the first and higher-order functional assemblies. The cost of the unit processes that go into making possible the increased value-added contributes to the added value obtained.

Table 4.4-1 illustrates a number of complex functional assemblies (systems) in increasing order of complexity. It serves to indicate that, at each higher stage of value-added, the resources involved become larger and the importance of successful and continued operation increases. Because of the size and financial factors involved, the number of competent suppliers and customers usually decreases as each stage of value-added is reached. Hence the greater is the departure from the classical concept of an unlimited number of suppliers and purchasers.

Another observation that has been made is that many user-customers, especially the government, tend to buy at the highest stage of value-added at which they can find competent suppliers. At these stages, the value of the system tends to be judged more in terms of its worth to the customer than of the intrinsic value of the material alone which performs the service.

4.5 Effect of System Time Phase on Criteria for Judgment

The process of judging may be performed at various times during the life of the system, and it is interesting to note that the bases of evaluation may change during this period. As noted in *Systems Engineering Tools*[49] in Chapter 3 on modeling, the accuracy of knowledge about the system requirements and equipment varies during the time period of the design and manufacture process. It is not surprising then that the judgment criteria may also change.

The three phases of system life which will be mentioned briefly in so far as they are influenced by judgment factors are the proposal phase, the development and design phase, and the acceptance phase after delivery of prototype or production systems.

Proposal Phase

During the proposal or conceptual phase, there exists the need for a comprehensive, broad judgment of the system in which the performance objectives and the principal method for obtaining them are set forth. A plan of system organization, as well as an outline for performing the overall job to accomplish the desired results, must be drawn up.

It is not generally possible to specify in detail the system configuration which will actually be employed because the information required to do the system engineering is not available and the cost and time involved to make such an evaluation would be quite significant. However, keeping in mind the admonition "sometimes in error, but never in doubt," it is essential to outline one principal approach which emphasizes the factors considered to be of greatest importance to the customer, i.e., high reliability, low cost, efficient performance, quick delivery, high novelty or innovation. It is worthwhile to indicate places where promising alternatives will be considered if the situation warrants.

Cost and time figures should emphasize the initial period of the system study as well as the design and development phases if these are required. Production costs and times cannot generally be based on facts at this stage and would probably be somewhat of a damper on the enthusiasm of the system's proponents if they were correctly stated.

The value of the system should be stressed, but since this factor frequently depends on the performance objectives, a high degree of confidence cannot be placed on its attainment.

Considerable importance must be placed on the method for performing the systems job and the people involved in doing so. The general basis for conducting the project should be stated clearly with a definite assurance that high company management is giving personal interest and attention to the undertaking if this is in fact the case. The extent to which the job will be done principally by the one company involved, or will be a joint responsibility with several subcontractors, must also be clearly stated. Here again the nature of the system will influence what decision seems most appropriate.

As the importance and worth of the proposal-phase effort to a system are becoming appreciated, more time, effort, and expense are being spent in sound and comprehensive activity at this stage.

Development and Design Phase

Judgment of the system during the development and design, i.e., definition and acquisition phases, stresses a continuous evaluation of the alternative configurations and actions with a high emphasis on the attainment of the performance objectives, including reliability, which were established initially. Parametric trade-off studies, optimization processes using modeling and simulation, are valuable in providing a sound basis for system judgment.

Design reviews as well as periodic time and cost reviews are helpful for recording progress as well as noting any significant changes from the original goals and objectives which appear justified on the basis of the more complete data and experience obtained. Where possible periodic projections should be made of time, cost, and performance figures to the end of the project.

Acceptance Phase Following Delivery of Prototype or Production Systems

At the acceptance stage, i.e., acquisition phase, the actual equipment has been shipped to the customer, and the system is very close to the point where it must speak for itself. Although one may say, "By their deeds you shall judge them," sometimes the methods available for proving satisfactory systems performance are by no means simple or even feasible short of inferential judgment.

Evaluation may be made either by the contractor alone, by the customer with the contractor's assistance, or by the customer alone. Frequently, it is necessary to perform a number of simulated tests under situations that check not all of the system's performance capability but enough of it to ensure a reasonable probability of satisfactory operation.

The system evaluations performed at this time form the basis for judging whether the contract requirements have been fulfilled or for the subsequent purchase of additional production systems. Hence these evaluations are extremely important and should have the benefit of thorough planning and execution by appropriately qualified personnel. Time and cost judgments can now be made with a high degree of certainty, and frequently little can be done except to increase the original allocations.

4.6 System Value as a Function of Performance, Cost, Time, and Reliability

Since value is an important criterion for judging a system, it is of interest to consider in a qualitative fashion how the value may be affected by some of the other major factors used in evaluation. Although the inference from the summation process indicated in block diagrams such as Figure 1.1-2 has been that performance, time, cost, and reliability contribute to the overall system value in an additive or linear fashion, the relationship of the value to these factors is frequently quite nonlinear. These nonlinearities may be different

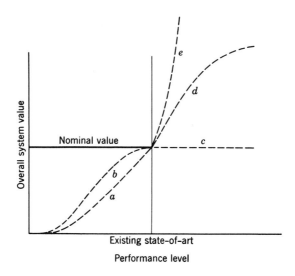

Figure 4.6-1. Various possible characteristics of overall system value versus system performance level.

from one system to another; the purpose of this discussion is to emphasize the need for endeavoring to establish the nature of the many value trade-offs which exist, both in fact and in the mind of the user or purchaser.

Value-Performance

If one considers the nominal value of a system to be the value associated with the performance level corresponding to the existing state of the art (see Figure 4.6-1), one notes that the values for performance levels below and above this standard may cover a large range. Thus, curve *a* shows that the value of such a system may fall off fairly rapidly with decreasing performance and may approach zero for a performance which is less than a third of the state of the art. Here the existing performance is marginal, and a slight decrease in its level causes the overall system value to fall markedly. Curve *b* shows a condition in which the value falls off from the nominal very slowly at first with decreased performance; however, for a sizable departure in performance, say below three-fourths of the existing state of the art, the value again falls off rapidly and the overall system value may again be virtually nil. Curve *b* is indicative of a kind of system in which the value of the overall system is initially dependent on some other system for its resultant performance. Hence, if this system's performance falls low enough, the overall objective may be realized only marginally.

For performance superior to the existing state of the art the value to the purchaser may be no more than for the nominal performance, curve c; may be proportional for a while and then level off, curve d; or may be steeply proportional to a very high value, curve e. These curves might correspond to conditions under which the existing state of the art for the system is (1) already such that it is no limitation to the overall system, curve c, (2) somewhat limiting but soon to be replaced by another limiting system so that further major improvements in this system would not be more valuable, curve d, or (3) extremely limiting to the overall system so that improvements in the performance of this system increase the overall system value way out of proportion to the improvement in the particular system performance, curve e.

Although for small perturbations in performance about the existing state of the art the increments in value may appear to be linear, in reality the nonlinearities may be large and may have to be taken into account.

Value-Time

Figure 4.6-2 sketches several value-time relationships that may exist if the system is made available earlier or later than the nominal

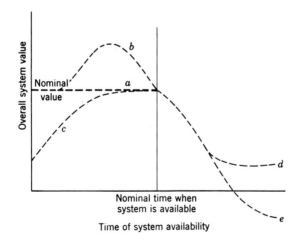

Figure 4.6-2. Various possible characteristics of overall system value versus time of system availability.

time. Curve a shows the case where obtaining the system early is no advantage or disadvantage. Curve b shows the situation where some benefit will be realized if the system is available a little while before the nominal time, but nothing is gained if it is received much before the original date and additional costs for storage may have to be incurred to handle its early arrival. Curve c again shows the case where early arrival of the system can be expensive and no corresponding benefit is realized.

Presumably, late availability of a system causes additional expenses or losses to be incurred that reduce the value of the overall system. Typical of such expenses are added labor charges for installation, personnel and interest charges on capital funds tied up in a nonproductive system. Curve d shows a case in which a lower limit to the value is reached because of late availability of the system. Curve e indicates that under some circumstances penalty clauses or other unscheduled expenses may be incurred which may cause the system value to become negative. In this event not only is the system value not positive, but also further losses may result.

Value-Cost

Value-cost curves, as shown on Figure 4.6-3, tend to have an intrinsically negative slope. For systems in which the cost is a comparatively minor consideration, the value is relatively insensitive to cost; see curve a-a'. In cases where the cost of the system is high relative to other costs, or where the lowest bidder receives the contract,

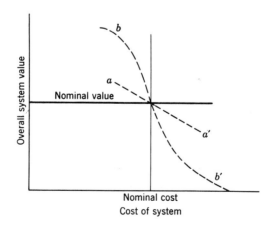

Figure 4.6-3. Various possible characteristics of overall system value versus cost of system.

System Value as a Function of Performance, Cost, Time, and Reliability 177

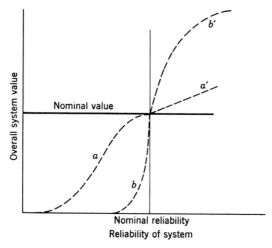

Figure 4.6-4. Various possible characteristics of overall system value versus reliability of system.

the value of the system may be strongly dependent on the cost; see curve b-b'. Presumably income factors, as discussed in Section 4.4, may also have some value-cost implications. For example, in the case of inferior goods a higher cost may make a system appear to be more valuable by taking it out of the inferior-goods category. Knowledge of the customer's value-cost characteristics for a system is quite useful in trying to establish the order of magnitude of an effort to embark on.

Value-Reliability

A small change in a part of a system may increase the reliability of that part of the system and therefore of the whole system, if that part is the limiting one. Hence small changes in reliability can represent the difference between mission success and failure. The value of a successful system versus an unsuccessful one may be very large. Figure 4.6-4 shows two typical curves of the value-reliability function. Curve a-a' indicates an inherently conservatively designed system in which the nominal value will be realized for nominal reliability; no great improvement in value will result from an appreciable increase in reliability, and some decrease in reliability can be tolerated without much loss of value of the system. Further decreases in reliability may reduce the system value appreciably, however.

Curve b-b' shows a marginally designed system in which a small decrease in reliability can cause a loss of the whole value. Added reliability over nominal may result in an overall system value in excess of nominal, say b', or may merely mean that the nominal value of the mission is achieved, as in a'.

The costs required to reach some desired level of system performance, cost, time, or reliability may be appreciable and significant. It is highly desirable to appraise oneself of the corresponding value of the overall system having these performance, cost, time, and reliability attributes. Armed with this knowledge, one is in a better position to establish the proper system trade-offs.

4.7 Conclusions

Many different factors are involved in judging the value of a system. Furthermore, there is a dependence of the system value on the particular time when it is determined. In many cases the value is strongly dependent on the extent to which system requirements properly reflect the user's needs. In addition to technical and performance requirements, such economic factors as the total net return from the system frequently provide a sound basis on which the system may be judged.

There are a number of different ways of considering the value of a system. Included in these are the traditional demand-supply curves as well as the demand-income and the value-added approaches. Each of these may provide a similar answer; when this is not the case, the system may be more satisfactorily sold on the basis of one method than of the others. Finally, even as the net return from a system will vary with time, the judgment criteria may have to be performed on different bases as the time phase of the system's life changes. It is axiomatic that in general only those systems which appear to have significant value are allowed to be started and therefore stand a chance of being completed. The sensitivity of system value as a function of such characteristics as performance, cost, time, and reliability may itself vary widely, depending on the environment in which the system exists.

5

Cost

5.0 Introduction

Cost is one of the most important factors in judging an engineering activity. In any society, and especially one in which the profit motive exists, the stated cost for accomplishing a given result is an easily measurable criterion by which the purchaser of the goods or service may compare the alternative methods and/or organizations proposed to perform the particular task.[56]

Although in fact the various systems proposed, if actually carried through, would not have comparable performance and/or could not be accomplished in the time initially agreed upon, these facts will in all probability not be provable at the time of awarding the contract for the job. However, the cost is something that remains the same, for a fixed-price job, from start to finish. It is generally established initially and adhered to. However, there are provisions in many contracts for altering the money received with detailed changes in the job scope according to some preset schedule, and sometimes bids are priced deliberately low with the realization that changes in job scope will permit increases in payment to compensate for an initial bid that is lower than would otherwise be justified.

Furthermore, transfer of funds in payment for a job represents a real interface between organizations. The purchaser has to obtain permission from some higher authority to buy the particular good or service, which must compete with other goods or services for the limited funds which the authority has available for its use. Cost is a real and a universally understood basis for comparison or trade-off.

To the organization quoting the cost of the job to be performed, this job in all probability represents only a portion of its overall manpower and resources. Consequently its total expenses, both fixed and variable, are not determined by the requirements of this job alone. The allocation of charges for this particular job may be made on more than one basis in terms of spreading certain costs, having broader value than for this one job, over a number of future possible jobs or other current jobs which may benefit also from the knowledge, skill, or facilities required for the particular job being costed. Thus, the method chosen to allocate costs may be open to certain subjective interpretation and judgment in the final analysis by the person responsible for making the quotation. Such conditions as the financial strength, position, and objectives of the organization costing the job are additional factors affecting the price to be quoted.

Another factor making for difficulty in establishing accurate cost figures for quotation purposes on systems engineering projects is the custom or one-of-a-kind nature of many of the systems involved. For these jobs, historical cost data are not available for the specific conditions, and furthermore the technical specifications may not be mutually understood by both parties. In addition, the novelty of the system to be built may be such that the actual technical requirements, in contrast to the contractual technical specifications, may be appreciably different from the prior jobs. From the nature of systems engineering, a considerable amount of design work may be required before the full extent of the job requirements, and therefore the job cost, are known. Thus neither the extent of the job to be done nor the cost for doing it may be known initially with a high degree of precision. It is small wonder that the *prices quoted* for systems engineering jobs may differ widely, as much as 3–5 to 1, for the "same" job requirement from different organizations.

Although the uncertainty and subjective values involved in costing systems engineering jobs will always be present to a degree, it is nevertheless necessary to establish a common basis for understanding the factors making up the total cost of a job. The costing components or factors are pretty much the same for all jobs, although the numbers assigned to them may differ appreciably. In the section that follows, we will describe methods for identifying the nature of the work to be done for the required goods or service and will mention its value to the customer. Later, the division of costs into fixed and variable will be discussed, and the problems associated with costs on a long-, medium-, and short-term basis will be presented. Finally, an example of cost-benefit analysis (Net return = Value of product or service—Cost) will be described in some detail.

5.1 What Is the Product or Service Being Costed?

Axiomatically, one of the major factors determining the cost of a systems engineering job is the nature of the product or service

What Is the Product or Service Being Costed

to be performed and the degree of thoroughness or intensity with which it will be accomplished. Figure 5.1-1 shows in a matrix-array form three combinations of systems engineering thoroughness and complexity that might be performed in a systems project.

The rows of the array describe the different phases of system engineering to be done for which cost figures may be desired. In each case it is necessary for some *study* to be performed. It is assumed that some sort of hardware or demonstration will be required. For a *breadboard* version, the demonstrable evidence may even take the form of a model or simulation. A *prototype* will also include some of the same hardware as the actual system but may incorporate only those items which are unique or especially significant to the particular system. A *production* system in addition to the prototype version also requires the complete equipment to perform in the actual environment for which the system is being designed.

The columns headed "Preliminary Design," "Intermediate Scale," and "Full Scale" in Figure 5.1-1 are used to differentiate the degree of completeness of the work to be performed. These terms in a very real sense are used to complement the breadboard, prototype, and production categories corresponding to the phases of systems engineering in the rows. Because of the magnitude of the sums of money, time, and effort associated with many large systems, it is found desirable to subdivide the financial commitments into these small, medium, and large categories designated by the three column titles. Referring to Figure 5.1-2, an analogy can be drawn between the organizational representation of the systems work to be performed, shown originally in Section 2.1, and the extent of completeness being considered here.

Roughly speaking, the system concept or the system design portion

Phase of System Engineering	Extent of Job Completeness		
	Preliminary Design	Intermediate Scale	Full Scale
Study	x	x	x
Breadboard	x	x	x
Prototype		x	x
Production			x

Figure 5.1-1. Phases of engineering work required as a function of completeness of system job to be done.

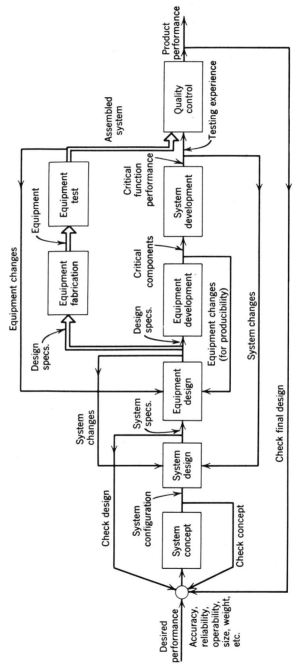

Figure 5.1-2. System engineering process.

What Is the Product or Service Being Costed

of Figure 5.1-2 would correspond to preliminary design as the term is used in Figure 5.1-1. Likewise "equipment design" or "equipment development" in Figure 5.1-2 would be included to correspond to the "intermediate scale" designation in Figure 5.1-1. "Full scale" as used in Figure 5.1-1 would correspond to the overall functions shown on Figure 5.1-2. In fact the costs for the full-scale systems effort in many cases will include not only the hardware operational system and its design as described in Section 2.1, but also the design, execution, and supply of the support system with its maintenance and training tasks. If, as outlined in Chapter 2, the overall systems job is far more extensive than the operational system itself, then the total systems cost in reality does include the support functions. However, for such purposes as accounting, cost control, or limiting the expenditures for any one appropriation, the costs for some features such as support may be omitted intentionally in the request for proposal (RFP).

The iterative nature of the systems approach shown in Figure 5.1-2 will, of course, tend to be reflected in the different cost figures, not only the total costs for each degree of completeness but also the absolute value of costs associated with each of the different phases. Figure 5.1-3 shows the ranges of total costs for each of the three extents of completeness as well as a total and relative breakdown

Phase of System Engineering	Preliminary	Intermediate Scale	Full Scale
		Total Costs	
	0.0001–0.001	0.005–0.20	1.0
Study	(30%) 0.0003	(10%) 0.0120	(3%) 0.03
Breadboard	(70%) 0.0007	(30%) 0.0160	(7%) 0.07
Prototype		(60%) 0.120	(30%) 0.3
Production			(60%) 0.60

Figure 5.1-3. Range of total and percentage costs.

for the costs associated with each of the four phases of the systems engineering effort. Although these ratios of the costs relative to the total full-scale costs will vary for different systems, the order of magnitude will tend to be roughly as shown. As can be seen from these numbers, the absolute magnitude of the study and breadboard work increases with the more complete job, although the percentage of the total effort for the particular phase goes down. From this it is quite evident that the costs to be quoted for study, as an example, are quite dependent on the overall magnitude and on the completeness of the systems work to be performed.

Not only is the extent of the systems job to be performed of significance, but also the detailed specifications of the performance, time, and reliability are important in providing a basis for arriving at a cost estimate. It is desirable that the customer state initially what his general and detailed requirements are. When such information is available, it should provide the basis for cost estimating.

However, in many cases the customer may not be able to specify his requirements in a definite fashion. In such cases, it is essential that the systems engineer describe in some detail *his understanding* of the systems work to be done and use this as a basis for reaching an agreement with the customer as to the product or service included in the cost figure submitted. In the event that this initial supplier-generated "specification" does not meet with the customer's approval, the difference in work items between what has been costed and what is actually required can be used to modify the original cost estimate. It is essential that the work requirements as well as the cost figures be understood by both customer and vendor at the time of initiation of the project. Since costs for training and maintenance may be as large as those for the operational equipment or even larger, whether or not these support charges are to be included should be definitely indicated.

5.2 Influence of Other Factors on Cost

The nature of the systems business is such that the effect of other systems may influence the costs of any particular system.

Common Facilities

In many cases the use of common facilities, equipment, or personnel may permit the costs to be shared over a broader base than would be possible if only one system were involved. Attention must be given to the possibility of effecting such economies by obtaining resources used only to a limited extent on a "buy" rather than "make" basis.

Comparable Costs for Compatible Systems

Furthermore, the cost of a given system should be influenced by the costs of the other major systems with which it is compatible. Thus, if the system is to work with systems that have long-life, high-reliability components, it is reasonable that this system be costed on a roughly comparable basis in the absence of specific information to the contrary. It is also reasonable that items priced very much higher than the average system equipment will receive a great deal more scrutiny and inspection than will items costing a great deal less than roughly comparable equipment.

Comparable Costs for Alternative Systems

Likewise, the costs must be judged on the basis of the costs for alternative ways of accomplishing the comparable function. From studies of the different technical ways of achieving a given set of customer's objectives, it is possible to provide alternative bases for making more than one cost estimate. This approach to preparing costs in effect represents a cost-sensitivity analysis. One aspect of this consists of preparing costs for various degrees of completeness or complexity of a particular method of achieving a given result; see Table 5.2-1. Another aspect of the cost-sensitivity picture is to prepare costs for the same degree of completeness or complexity for several alternative methods of achieving approximately comparable results; see Table 5.2-2. In some cases, it may be worthwhile to

Table 5.2-1. Costs (Dollars) for Various Systems Using Method I

	System Capabilities		
Method	Minimum Acceptable System	Adequate System	Complete System
System I (electric)	15,000	20,000	23,000

Table 5.2-2. Costs (Dollars) for Comparable Performance Using Various Methods

	System Capabilities		
Method	Minimum Acceptable System	Adequate System	Complete System
System I (electric)	15,000	20,000	23,000
System II (hydraulic)		25,000	
System III (pneumatic)		18,000	

explore the other degrees of completeness for the other methods of accomplishing the systems objectives.

In any event, it is highly desirable to have explored the costs for a number of different variations about the so-called "nominal" system which has been asked for by the customer or appears necessary for the job. Having these alternative figures may convince the user that your estimate for the method you recommend or the one he specified is the cost he is willing to pay, or it may encourage him to change his requirements to bring them into line with a cost that he can afford.

Elasticity of the Market Demand

Since some items of cost are relatively insensitive to changes in the number of systems to be produced, while others are influenced to quite an extent by the size of the market demand, it is worthwhile to explore the effect of possible additional quantities on the cost for the initial lot size as well as for follow-on orders. With an unlimited (inelastic) market demand, the unit system price is fixed, and the income from later systems will be much the same as from the first one. With a limited (elastic) market demand, the unit system price may fall off rapidly, and the income from later systems may be much less than from the first one. Depending on the elasticity or inelasticity of the demand for the later systems, as well as the possibility of selling later systems, the cost (i.e., price) of the initial systems may be modified to incorporate the influence of possible future orders.

5.3 Long-, Medium-, and Short-Term Cost Considerations

There are many different sorts of decisions which affect the cost of a system over its life. Some of these decisions strongly influence the system's long-term capabilities or capacity. For example, the installed plant capacity of a system will determine the load it can supply and therefore serves to provide somewhat of an upper limit to the magnitude of the return which can be realized for a reasonable period of time in the future. Although it may be that *less* production will be realized from the system because of lack of demand or inability of the equipment to operate, it is not likely that much *more* production can be obtained than that for which the system is designed. Although as much as 20–50 per cent more output than given by design values can sometimes be realized, significantly greater outputs will probably require additional capacity and cost.

Given a plant or system capable of a certain maximum production and faced with the job of supplying a certain demand schedule for a period of time, there are various ways of using the installed plant and the associated variable resources for which the total costs may be quite different.[17] The time intervals involved here are called medium, in that they are short enough that even the expenditures for additional long-term costs would not appreciably increase the amount of productive output that would be available in the time period of interest, and they are long enough so that shorter-term changes in load demands or equipment failures may make it necessary for the medium-term decisions to be altered quickly as the actual demand schedule for load as a function of time develops.

Costs associated with short-term decisions are ones influenced by unexpected events of brief duration which permit a given plant of a certain long-term size, scheduled to operate on a certain medium-range basis in accord with a given estimate of load demand (or resource availability), to operate actually in a short-time manner which is most propitious for the existing events. Overload detectors, overspeed- and acceleration-limiting devices, load-frequency regulators, overtemperature indicators, and over- or undervoltage relays are typical of the devices and means that are associated with short-term costs. Failure to make short-term decisions promptly and correctly may result in plant and/or equipment failures that can be very costly to repair or an inability to produce the desired product and/or service.

Figure 5.3-1 shows in matrix form how the time scale of the long-term cost decisions covers primarily those years in which it is assumed

Cost Factors Dominant	Decisions Covering Years	Decisions Covering Days, Weeks, and Months	Decisions Covering Seconds, Minutes, and Hours
Long term	x	Current practice	Current practice
Medium term	Fixed	x	Current practice
Short term	Fixed	Fixed	x

Figure 5.3-1. Time span of cost decision influence.

that the medium- and short-time decisions are made in the appropriate current fashion. Likewise, the medium-term decisions are shown to cover days, weeks, and perhaps months in which the long-term decisions are essentially fixed, and the short-term decisions are again assumed to be made in accordance with the most effective current practice. Short-term decisions are shown to be ones of importance over seconds, minutes, and perhaps hours in which the long-term and medium-term decisions are essentially fixed. Figure 5.3-2 shows in

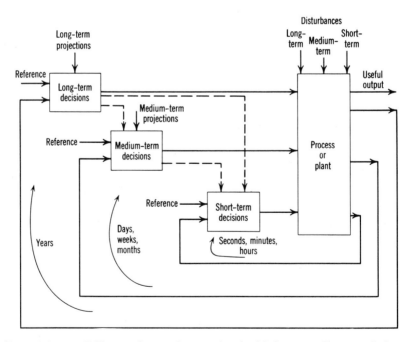

Figure 5.3-2. Different time scales associated with long-, medium-, and short-time decisions and their control of a process or plant.

a block diagram fashion how growth projections and external disturbances each influence the three types of decisions and how each decision process provides inputs to the process or plant. The dotted lines indicate that the longer-term decisions (i.e., higher-level ones) may also be used to modify the shorter-term decision processes themselves, in addition to operating on the plant in their long-term role.

During the long-term decision process, frequently called systems planning, alternative plans for system growth to meet anticipated load needs are studied. Costs for plant, investment financing, operations, etc., are all considered for various growth plans and plant and equipment expansions. Transportation systems, utility systems, housing developments, manufacturing plants, and communication systems are typical of the organizations which perform cost analyses of this type.

The medium-term decision processes affecting costs are designated by such terms as scheduling, programing, and inventory control, and involve ways of utilizing the existing plant, manpower, and material resources most effectively in an organized fashion to obtain a favorable cost control situation. Decision rules for buying materials in most economical lot sizes, at most favorable times as far as market prices are concerned, and in such ways as to hold inventory costs at a proper level for the production needs, are representative of cost control decisions performed on a medium-term basis. Periodic maintenance plans for balancing the cost of preventive maintenance against the probable costs for failure and shutdown if such maintenance is not performed are also typical of medium-term cost considerations. Military as well as industrial, utility, and transportation systems all give consideration to such medium-term cost decision processes.

Short-time cost consideration processes are generally employed as part of the regular on-line automatic controls of the system. These controls may improve the quality of the product and thereby justify their cost by the increased system output value they engender. They may increase the useful output of the system, so that their cost can be justified on the basis of a greater total value for the product. They may increase the reliability or safety of the system on a continuing basis and therefore be useful in creating more system output than their cost requires. In general, the use of such short-term decision methods is feasible on the basis of current data, and extensive speculation or estimation for determining future costs is not necessary.

The short-, medium-, and long-term cost considerations may involve quite different time periods over which their effectiveness is pertinent. They also tend to involve people with as varied training as marketing, financial, and business for the long-term costs; marketing, production, and design for the medium-term costs; and production and engineering for the short-term costs. Correct decisions in all three areas are important if the greatest overall cost effectiveness of the system is to be realized.

5.4 Build-up of Total Cost from Cost of Parts

The preceding sections have stressed the importance of preparing cost estimates based on the nature of the product or service desired by the customer and the influence of other factors, such as quantity of systems to be supplied and time periods involved. Once having established the particular framework within which the costs are to be determined, it is necessary to subdivide the costing structure into a number of parts within which separate figures can be drawn up.[18, 23] In this way, there is developed a method for establishing cost goals for the various parts of the work, and therefore a means for obtaining references against which actual costs on the job can later be compared. Furthermore, by subdividing the costing by parts or tasks, the possibility of error in the total figure is reduced because the errors in the individual parts of the estimates tend partially to cancel each other.

Of the various different cost structures which can be employed, a convenient one is that shown in Figure 5.4-1, where the time phases of the job and the fixed, variable, and other charges and elements of cost appear as coordinates for the matrix. Since equipment is implied in the description of the time phases, such a matrix array will be required for each major equipment item as well as for the systems integration items that may be needed. The terms fixed costs and variable costs are used to indicate respectively the costs which

Major Cost Components	Time Phases			
	Study	Breadboard	Prototype	Production
Fixed costs		x	x	x
Variable costs	x	x	x	x
Past investment and other charges			x	x

Figure 5.4-1. Allocation of major cost components as a function of different time phases.

are relatively independent of the specific use of this particular system, i.e., fixed, and those which are quite strongly dependent on the use of the system, i.e., variable.

The fixed and variable cost concept is most useful in connection with systems producing hardware; for systems-studies types of activity the fixed costs tend to get lumped into an overhead percentage which is assigned to mark up the variable costs associated with the direct labor. It will be noted that the general and miscellaneous expenses, and the maintenance and operating expenses of "fixed charges" to be described below, are among the principal costs associated with the overhead percentage that is added to direct labor or the variable expenses normally considered. One of the problems associated with the concept of adding an overhead percentage to direct labor charges is that, when the amount of direct labor expended overall departs significantly from the estimated annual figures, it will be necessary to obtain the money to cover the fixed costs by using a different overhead percentage figure.

Cost Basis for Initial System Investment

The problem of determining costs for an overall system investment has been present for a long time and must be faced any time one makes a major capital expenditure.[23] A number of companies,[3] especially electric utilities,[8, 19] have been concerned with this problem to the point that several well-thought-out and well-organized approaches have been developed for proceeding with the economical comparison of alternative facilities. Many of these principles may be applied to advantage in the case of systems for which appreciable capital expenditures are required.

As a convenient way of obtaining the overall cost factors in a major system investment, the following types of charges must be considered.

A. *Fixed charges.* These costs are dependent on the magnitude of the initial capital investment and are relatively insensitive to the method of operating the equipment.

B. *Variable charges.* These costs are dependent on the method of operating the plant but are also influenced by the characteristics of the plant as well as the level of output or yield.

C. *Past investment and other charges.* These costs are dependent on factors that are essentially outside the control of the manager of the operation but nevertheless alter the charges for the facilities or process.

Each of these types of charges will now be investigated in more detail.

Fixed Charges. For the purpose of determining costs for the initial system investment, the following factors have been shown to be significant in considering fixed charges:

1. Return on investment.
2. Depreciation.
3. Federal income taxes.
4. State income taxes.
5. Property taxes, other taxes, and insurance.
6. General and miscellaneous expenses.
7. Maintenance and operating expenses.

The magnitude of the fixed charges is dependent to a large extent on the initial capital investment or "first cost," where first cost is the term applied to the money required to build the plant and is sometimes referred to as capital expenditure or plant investment. Investment in plant covers the original cost of such plant, taking into account all associated expenditures charged to the plant accounts up to the time the project is completed and ready for use. Among the items included are the following:

Material: all material used or consumed in the construction of the plan plus associated freight; sales tax if applicable, and supply expense.

Installation: all direct labor and motor vehicle costs and incidental expenses.

Miscellaneous loadings: costs for supervision and the use of tools, general expense, social security taxes, and relief and pensions.

Engineering: engineering time and associated costs.

Other charges during construction: interest during construction, taxes, insurance where necessary, etc.

After the magnitude of the first costs has been established for alternative proposed facilities, the fixed charges associated with these first costs can be determined. The following categories are the major ones.

RETURN ON INVESTMENT. This charge represents the cost for the use of the capital invested in the plant or equipment. Depending on some such criterion as current or anticipated interest rate for money that might be borrowed for this purpose, or the rate of return that could be obtained from an alternative project investment, the value of the rate of return on investment can be selected. This rate times the appropriate portion of the total first cost determines the magnitude of the cost for "return on investment." Figures of from 6 to 10 per cent per year are representative of the range of values used for this charge.

DEPRECIATION. The depreciation charge is made to cover the reduction in the value of the equipment that occurs as a function of time rather than use. Depreciation is influenced by such factors as average life, including the effect of obsolesence, and the ultimate net salvage. All of these factors must be estimated and therefore are matters of judgment. In addition, when capital has been borrowed to finance the purchase of the system, it may be necessary to use the depreciation charge as a means of raising the sums of money required to repay this purchase loan.

The charge for depreciation on a new plant is, in many cases, the most difficult to evaluate of any of the items of annual cost in engineering studies. The annual cost for depreciation in such studies is based on the expected service life of the particular equipment being considered. On the other hand, accounting practices frequently base the accruals to depreciation reserve on the estimated average service life of a group of equipment items. For purposes of cost studies it is necessary to consider carefully the service life expectation for the particular equipment and not necessarily use "average" values based on accounting data.

FEDERAL INCOME TAXES. Federal income taxes are annual charges that are levied on taxable income as defined by applicable laws. They are functions of the income, the income tax rate (about 52 per cent at present), and the method of accounting for depreciation.

The following formulas express the basic steps involved in income tax calculation:

Income tax = Taxable income × Tax rate
Taxable income = Gross income before income tax
− Interest on debt
Interest on debt = (Initial investment − Depreciation to date)
× Rate of return on investment

Although the federal income tax is a function of such fixed charges as initial investment and depreciation, it is also a strong function of gross income and is therefore dependent on external factors, i.e., market demand, as well.

STATE INCOME TAXES. State income taxes may be calculated as for federal income taxes after ascertaining the statutory definition of taxable income and the applicable state income tax rate. The nature of the depreciation schedule or accounting may also differ in computing the state income tax. Where state income taxes exist, they may be deducted from the gross income for purposes of calculating the federal income tax. Depending on the state, the federal income tax may or may not be deductible in calculating taxable income for the state income tax.

PROPERTY TAXES, OTHER TAXES, AND INSURANCE. Property taxes are usually based on the assessed value of the plant. Where comparable equipment or property exists, it is possible to estimate such costs by including information on present and anticipated property tax rates. Other taxes, such as utilities, vary from community to community and must be included as applicable for the particular area of interest.

The cost of providing insurance, when incurred, is treated as an annual cost of doing business. Some large companies usually do not insure their property; instead provision is made for casualty losses depreciation of general and miscellaneous expenses.

GENERAL AND MISCELLANEOUS EXPENSES. These include salaries of executives, general office supplies and expenses; legal, accounting, and marketing services; welfare and pensions, etc. Such expenses are frequently expressed as a fixed percentage of gross income or plant initial investment. In some comparative engineering studies these expenses do not differ between alternative facilities.

MAINTENANCE AND OPERATING EXPENSES. Current maintenance expenses include the cost of ordinary repairs and day-by-day upkeep of equipment. The costs of such items as troubleshooting and repairs, inspections, and routine tests and maintenance fall into this category. The costs of replacing minor items of the plant, including labor and materials, are also part of maintenance expense.

Maintenance expense may be estimated as a percentage of first cost, or it may be expressed as a specific number of dollars per year for the particular project or equipment under consideration. When significant changes in maintenance or repair requirements may result from different ways of operating the plant, these expenses should probably be considered as part of the variable costs.

Variable Charges. Variable charges are ones that are incurred as a result of the operation of the equipment or plant and may differ depending on the rules or strategies for its operation. These costs are also a function of the product output of the plant. Significant

Build-up of Total Cost from Cost of Parts

components of variable costs include:

1. Charges for input materials.
2. Cost of processing or preparation of the final output.
3. Value of scrap or waste materials.
4. Miscellaneous expenses that are a function of output.

The *input materials* are those which are the necessary inputs to make the products desired and which are a function of the level of output. Included are the costs not only of the direct input materials that show up as part of the output, but also of the catalyst type of materials, as well as auxiliary charges for utilities, heat, cooling costs, etc., which are necessary to obtain the desired output. These costs must be expressed as a function of the product output, as well as any environmental factors that are of significance. Average or weighted values for these environmental factors, based on reasonable estimates over the periods of time under consideration, should be used.

Costs of *processing* or *preparation of the final output* represent those associated with taking the output from one process and bringing it into the condition necessary for its ultimate use. In some cases the output of the process may have to be shipped to the user, as, for example, the transmission of hydroelectric energy from power plant to load center. These costs again should be expressed in terms of the output as well as any other environmental factors that may be pertinent. In other cases the qualities of the process output may not be such that it can be used directly, and additional preparatory expenditures which are a function of the magnitude of the output are required.

Value of scrap or waste material covers the salvage value or other return realizable for the output that cannot be sold as first-grade product. These sums are essentially negative costs in that they represent a useful return, and they should be included as a variable cost if in fact they can be obtained. For example, some scrap metals have value, and off-standard materials can sometime be reworked satisfactorily and sold.

Miscellaneous expenses that are a function of the output cover the general expenses that are otherwise unassigned but are related to the magnitude of the output. Unusual maintenance expenses that are incurred as a result of exceeding output ratings are an example of such a cost.

Past Investment and Other Charges. Past investment refers to the expenses associated with old equipment that will be replaced in function by the new equipment. To the extent that certain fixed charges remain on the old item when it is retired, these charges must be met regardless of which alternative of the other methods is employed. The salvage value of the past investment represents a negative cost which will be available to help reduce the cost of the new equipment. Salvage value may be considered as the market value that can be realized by a sale or by the value for reuse of equipment if it can be employed elsewhere.

Another factor which may be handled as an "other charge" is the cost associated with price level changes. Projection of price level changes is a hazardous undertaking, and such estimates should be used cautiously.

The preceding subdivision of costs into three major factors, with further subdivision of each of these factors into a number of separate components, demonstrates that the total cost can be built up as the sum of its parts. The necessity for repeated interations is evident from the percentage nature of a number of the individual cost terms which are dependent on factors that may not be known initially, for example, gross return or maintenance and operating expenses. The use of literal expressions for some of these type terms may permit an analytical form for these relationships; if this is not possible repeated iterations will generally converge quite rapidly.

In this discussion the problem of costing an initial system investment has not stressed all of the time phases illustrated in Figure 5.4-1; the method here may be considered as being directed more particularly to the total cost picture from study through production.

Example: Cost Comparison Study for Revising Motor Line Design

The preceding material describing the factors that should be considered in an initial system investment study has been largely influenced by experience in making comparisons of costs for alternative facilities in the electric utility field, where a large number of detailed examples may be found in the technical literature.[8, 19] To provide another illustration of how long-term cost considerations can be used in other applications and in a preliminary fashion to yield worthwhile results, the following example from the motor design field will be employed. Since only a comparison of alternative design policies is being considered, a number of cost factors are neglected in this preliminary consideration and only a few significant relationships are included.

Build-up of Total Cost from Cost of Parts 197

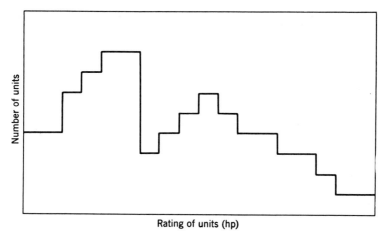

Figure 5.4-2. Market estimation of number of units versus rating size.

Problem Statement: A market survey has been made as shown in Figure 5.4-2, which indicates the number of units that can be sold as a function of the size of the units for a given number of poles (i.e., speed). A detailed redesign of an existing motor line has developed a family of curves of costs based on the assumption that six different-sized motor punchings should be used to cover the range of sizes required to meet the above market. It has been suggested that an economy can be realized by making a fewer number of diameter models and allowing a greater length of machine for each diameter, i.e., increasing the maximum allowable L/D. Additional development and tooling costs will be required to extend the maximum L/D ratio, but it is felt that these costs will be less than would be required to make the tools for one or more additional punching-diameter sizes.

The major cost factors to be considered are the following:

1. FIXED CHARGES. The fixed charges are attributable to *capital costs for tooling* and to *engineering costs for development*. It is assumed that these can be depreciated linearly with time, for example, the tooling over a 4-year period and engineering over an 8-year period. Taxes and other fixed expenses are assumed to be comparable for the different number of rotor punchings as a first approximation. Also return on investment is neglected in the initial step.

Detailed estimates have indicated that the cost of tooling of a given punching diameter as one extends the L/D span of the ratings in the diameter (adds more ratings) is as shown in Figure 5.4-3.

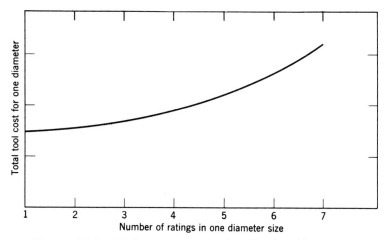

Figure 5.4-3. Tool costs versus number of ratings/diameter.

Likewise, estimates have shown that the cost of development of a given punching diameter as the L/D span of the ratings in the diameter is as shown in Figure 5.4-4.

2. VARIABLE COSTS. The principal variable costs that will be sig-

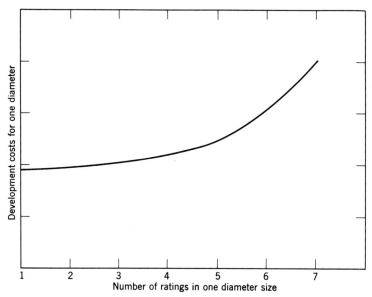

Figure 5.4-4. Development costs versus number of ratings/diameter.

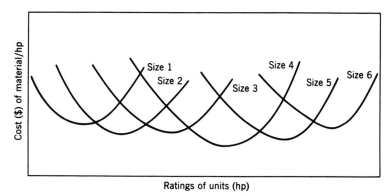

Figure 5.4-5. Cost of material versus horsepower for various sizes of punchings.

nificant are the *cost of material* and the *cost of labor*. Other variable costs, such as those involved in shipping and other miscellaneous processes, are considered to be the same for each method.

The *cost of material* for a machine of given horsepower as longer machines are built in a constant punching diameter is shown in Figure 5.4-5 for the design using six punching diameters. Labor is considered as 20 per cent of the material cost.

3. PAST INVESTMENT AND OTHER CHARGES. These costs are not affected by the selection of one or another of the choices available and will not be included.

ASSUMPTIONS AND COST FORMULAS. Other relationships which are significant in performing this study include the following:

A. The horsepower, hp, has been found to vary with diameter, D, and length, L, as

$$\text{hp} \cong D^a L^b, \text{ where } a \text{ and } b \text{ are known constants} \quad (5.4\text{-}1)$$

B. The first rating to be placed in any diameter punching will be at the point of minimum \$/hp material and labor, Figure 5.4-5, and then additional ratings are placed at either side of this point.

C. It is assumed that the tool and development expense are the same for any diameter punching.

D. It is also assumed that, for the number of poles being considered, the total tool and development expenses are prorated according to the units represented and included in this survey.

Although the overall answer to the problem of what is the proper number of punching diameters, will not be determined here, a start will be made to show the nature of the procedure for finding the cost of building a number of different horsepower sizes in a given punching diameter. The following cost formulas are appropriate:

$$\$/\text{hp} = \frac{\text{Cost tools/yr} + \text{Cost development/yr} + \text{Cost material and labor}}{\text{Total hp}} \quad (5.4\text{-}2)$$

$$\text{Cost tools/yr} = \frac{\text{Cost tools (Figure 5.4-3)} \times \text{Prorated units/Total units}}{4 \text{ yr}} \quad (5.4\text{-}3)$$

$$\text{Cost development/yr} = \frac{\text{Cost development (Figure 5.4-4)} \times \text{Prorated units/Total units}}{8 \text{ yr}} \quad (5.4\text{-}4)$$

Cost material and labor
$$= 1.2 \ \$/\text{hp material for specific punching (Figure 5.4-5)} \\ \times \text{Units sold (Figure 5.4-2)} \quad (5.4\text{-}5)$$

$$\text{Total hp} = \Sigma(\text{hp}_1 \times \text{Units}_1 \text{ sold} + \text{hp}_2 \\ \times \text{Units}_2 \text{ sold} + \cdots) \quad (5.4\text{-}6)$$

Using the point of minimum $/hp on the third curve from the right as a starting point, one calculates the $/hp for building first this size of machine, then for this size and the two adjacent sizes of machines, then for these three sizes and the two adjacent sizes of machines, etc., until the $/hp increases. Figure 5.4-6 shows the results of such calculations and indicates the cost to be a minimum when five machines are built with one punching diameter.

At this point, no more ratings should be included since the cost of all ratings in this diameter has been minimized. The next largest rating should then fall in another diameter above the one just considered, and the next smallest rating should fall in a diameter below the one just considered. Using information regarding maximum and minimum L/D, the hp-D-L relationship of Equation 5.4-1, and the material cost-size data of Figure 5.4-5, one can construct similar $/hp curves for other punching sizes. In this fashion, the cost equations can provide the overall data from which decisions of system investment costs can be made.

The above illustration of system investment costs has been purposefully simplified; it is of course evident that succesive computations involving more detailed cost figures and the inclusion of additional terms that have been omitted initially may be required before final decisions can be made. However, the value of using cost equations in helping to arrive at these decisions has been demonstrated.

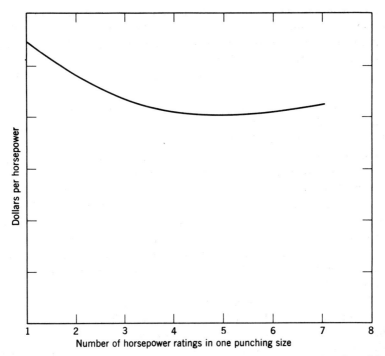

Figure 5.4-6. Dollars per horsepower versus number of ratings/punching diameter.

5.5 System Costs Considering Only Variable Costs

Another aspect of systems costs considerations which is of importance at times is the determination of the cost of operation of a system which is already in place, so that the initial investment costs are essentially not subject to change.[31] For these systems the problems are ones of trying to minimize the variable costs. The minimization may be such as to take place over a period of time during which the variable costs associated with the production or demand for a product are not constant with time, or over a period in which the demand for the product is independent of the time when the product is made or used. Each of these situations will be explored briefly since it represents an interesting facet of cost considerations from a systems viewpoint.

Medium-Term Cost Considerations

Medium-term cost considerations apply to the determination of costs in cases where the processes or equipments are already in place and are not to be changed, where the availability of raw materials or energy is limited, or where the product or service has different values depending on the time when it is available. The overall characteristics of the process have been determined by the initial design of the equipment, and therefore the fixed costs are not subject to change.

Because there may be a limited amount of energy or resources and the load requirements may change with time in an unpredictable fashion, it is not possible to program accurately in advance the optimum method for using the resources available. In general a program is prepared to handle the nominal conditions that are expected, with the medium-term cost considerations providing the principal means for control decisions. Because of the availability of alternative sources or kinds of materials that are capable of producing the desired products or service, that cost must continually be determined which will yield the greatest net return over the period of time of interest. This information provides the basis for continually modifying the basic input to a cost control.

Medium-term cost considerations may be viewed as helping to provide information with which cost control of a batch-type process may be performed on a predictive or optimizing basis.[49] In such an event, the model will be used repetitively with new input data as the events in the cost picture change with time. Figure 5.5-1 shows a representative product demand curve that varies with time. The demand for electric power generation per day or for chemical production over a longer period of time might be typical of the type being considered. The demand is such that it can be supplied by one or more units or equipments which have different production costs, different total amounts of product they can produce, different costs for making the equipment ready to produce, and different standby operating charges.

Although the most economical method of operating the one or more units can be preprogrammed from the estimated demand, as the situation develops the actual demand will probably differ from the estimated one. Hence it is desirable that equations for the variable cost be available to permit a recalculation of the program on the basis of present conditions and estimates of future demands. An element of prediction is involved in the estimate of future demand, and a control in which is incorporated some knowledge of the logic that will affect this demand will be better able to meet these future requirements. The variable costs must be considered over the period of time in the future during which alternative programs can be carried out. An optimizing control, in which the integral of the total variable

System Costs Considering Only Variable Costs

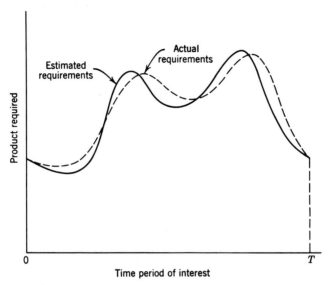

Figure 5.5-1. Estimated and actual product requirements versus time.

costs over the remainder of the operating cycle is the criterion, could be used for such a case. Thus, the control objective would be to make

$$\int_{t_1}^{T} (\text{Variable costs}) \, dt = \text{minimum} \tag{5.5-1}$$

The following represent items of costs that may be present for each of the alternative methods.

Input materials costs, including fuel costs based on the effectiveness of alternative methods as well as the cost of the various constituents necessary to make the desired output. In general, these will be a function of the level of output.

Start-up and shut-down costs, including material and time required to get equipment in condition to supply product of desired quality. Costs involved in shutting down the equipment should also be included in this factor.

Standby costs needed to provide reserve capacity: in the case of electric utility to meet demands in excess of estimated, or in the case of chemical processes to ensure adequate reliability back-up.

Penalty costs associated with the added expenses of obtaining the product at one place or another, e.g., additional electrical losses; or of refining the product more fully, as in the case of one method of chemical preparation or another; or of being forced to buy some of the product on the open market to meet an unexpected demand.

To round out the medium-term cost considerations, it is highly desirable that some additional probability data be available to indicate the chance that one or another of the alternative methods or unforeseen difficulties has of happening. Reliability information, which is also probabilistic, is likewise important in evaluating the medium-term costs. Referring to Equation 5.5-1 and to Figure 5.5-1, one notes that this problem is similar to the traditional one of dynamic programming or dynamic optimization.[49] Hence some existing methods are available for attacking this problem, although the job with nonlinear relationships is by no means one that is readily done without iteration.

Short-Term Cost Considerations

In considering short-term costs, one assumes that the long-term "fixed" charges are frozen and evaluates the expenses associated with the current values of the external environments and the controlled and manipulated variables of the process or facilities. Thus, the system gains associated with high product or performance quality, and the costs associated with raw material or utility expenses, are related to variables of the process or equipment. Although some of the short-term values of the controlled and/or manipulated variables may not result in any immediate expenses attributable to long-term failures, the long-term failure effects can be accounted for in terms of current "average-loss" rates that are expressed as functions of the controlled or manipulated variables.

Figure 5.5-2 shows in block form the characteristics of a process, its outputs, its process variables, its external environments, and its manipulated variables. These and the design parameters are the quantities which influence the control of the process, its value, and the cost associated with operating it.

The basic net return equation can be written as:

$$\begin{array}{c}\text{Total net} \\ \text{return per} \\ \text{unit of time}\end{array} = \begin{array}{c}\text{Total value of} \\ \text{products produced} \\ \text{per unit of time}\end{array} - \begin{array}{c}\text{Total cost of} \\ \text{producing} \\ \text{products per} \\ \text{unit of time}\end{array}$$

$$R_N \;\; = \;\; V_T \;\; - \;\; C_T \tag{5.5-2}$$

where the total value of the products, as well as the total cost of producing them, is indicated.

System Costs Considering Only Variable Costs

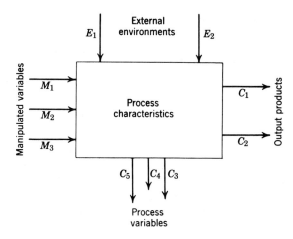

Figure 5.5-2. Block diagram of process with its inputs and outputs.

Presumably the same or comparable time intervals will be involved for each method considered; therefore all terms will be referred to on a per unit basis in the equations that follow.

Expanding the equation into its parts,

Total net return = (Value per pound of product) (Number of pounds produced) − Variable costs − Fixed costs

$$R_N = \left(\frac{V_p}{W_p}\right) \times W_p - C_V - C_F \qquad (5.5\text{-}3)$$

Further amplifying Equation 5.5-3 and regrouping terms,

$$\frac{\text{Total net return}}{+ \text{Fixed costs}} = \begin{pmatrix}\text{Value per pound} \\ \text{of product}\end{pmatrix}\begin{pmatrix}\text{Number of pounds} \\ \text{produced}\end{pmatrix}$$

− (Total variable cost of materials or utilities used
+ Variable cost of refining product made)

$$R_N + C_F = \left(\frac{V_p}{W_p}\right) W_p - \left[\left(\frac{C_m}{W_m}\right) W_m + \left(\frac{C_r}{W_r}\right) W_p\right] \qquad (5.5\text{-}4)$$

where W_m is the number of pounds or amount of materials or utilities required.

There are a number of ways of expressing the total product yield or number of pounds of product produced; the following method of describing this quantity will be used for purposes of illustration.

Number of pounds produced = Total energy input

$\times \dfrac{\text{Pounds produced}}{\text{Energy input}} \times$ Production efficiency

$\times$ Fraction of time operating

$$W_p = E_T \times \left(\frac{W_p}{E_p}\right) \times N_e \times F_t \qquad (5.5\text{-}5)$$

where *production efficiency*, N_e, is the fraction of theoretical conversion that takes place in the equipment or process for the present condition of operation. *Fraction of time operating*, F_t, is the time when the plant is operating at production efficiency, N_e, divided by the time when the plant is operating at this efficiency plus the time when the plant is not operating.

Before Equations 5.5-4 and 5.5-5, which summarize the cost relationships which exist for the process, can be used effectively, they must be related to the process variables, characteristics, and other quantities shown on Figure 5.5-2. Since the manipulated variables, the external environment, the process variables, and the output products are the physical quantities that can be measured and controlled, they must serve as inputs to the cost equations.

Although these detailed relationships will depend on the particular process or equipment being studied, the following general relationships are typical of those that may exist. Manipulated variables such as M_1, M_2, and M_3 will appear as portions of the variable cost of materials, W_m, in Equation 5.5-4. Process variables like C_3, C_4, C_5, and environmental factors E_1 and E_2, representing temperature, pressure, vibration, etc., will appear as terms in such factors as the production efficiency, N_e, and fraction of time operating, F_t, in Equation 5.5-5. The actual output products, such as C_1 and C_2, represent all or part of the pounds of product produced, W_p, appearing in Equations 5.5-4 and 5.5-5.

The detailed relationships between the various process characteristics and variables and the corresponding cost quantities must be developed over a period of time, although initial engineering and cost estimates of these relationships can and should be made. In addition, on the basis of actual operating experience, further improvements in the understanding of and values for these relationships can be made with use of the cost equations.

Presumably, the fixed costs, C_F, can be determined from long-term cost information that may be available. For some purposes where these fixed costs are not influenced by the operating conditions, they

System Costs Considering Only Variable Costs

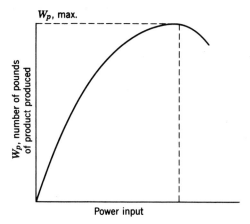

Figure 5.5-3. Pounds of product produced as a function of power input.

can be neglected because it is only the relative values of the net return R_N, which are of concern.

Example: To provide a means for interpreting cost equations 5.5-4 and 5.5-5, a graphical representation will be given to their terms in the following example. Since the energy input in Equation 5.5-5 is being considered per unit of time, the *energy input* term is in reality a power input and is used as such in the example.

The number of pounds of product produced per unit time, Equation 5.5-5, is a function of the power input. However, the relationship is not a linear one nor does it continually increase, as is shown in Figure 5.5-3. The falling off of the pounds of product produced per

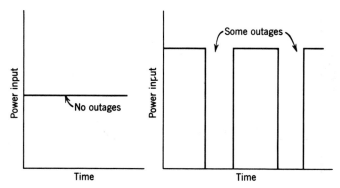

Figure 5.5-4. Power input versus time at two different power input levels, showing operation with and without outages.

unit time is a result of the increased number of equipment outages as the power input and the accompanying equipment stresses rise. This effect is shown in the two examples for Figure 5.5-4, one for a power input for which no outages occur, and the other for a higher input with a number of outages.

In determining the variable cost of material used, the *term W_m is taken to be proportional to the power input level,* so that this cost continues to increase its value when the maximum number of pounds produced is exceeded.

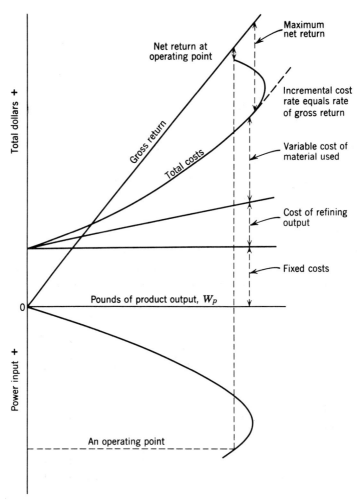

Figure 5.5-5. Composite cost and returns picture versus pounds of product output.

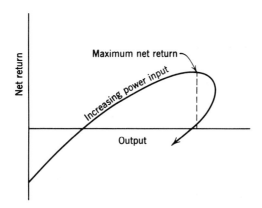

Figure 5.5-6. Net return versus output for increasing power input.

Figure 5.5-5 shows graphically the composite characteristics of costs and net returns as a function of the pounds of product output. Although the total costs exceed the gross return for low values of pounds of product output, a break-even value is reached after which the net return becomes positive for an increase in output. Still further increases in output can only be obtained up to a certain maximum value, after which the output becomes lower again. The maximum net return is obtained when the incremental rate for producing the output products is equal to the rate of yield (value) for the gross return. Further efforts at increasing the output above this point result in reduced net return immediately, and later cause an absolute reduction in output. Figure 5.5-6 shows these relationships.

In summary, the short-term cost considerations permit one to express in terms of the present process variables and characteristics the net return from the process or equipment, taking into account the average long-term reliability or operating efficiency effects based on design and/or previous experience. This is particularly helpful in costly, slowly changing, long-time operating processes where high reliability is essential but the value of increased operating effectiveness can be appreciable. It is also of great value in costly, rapidly changing processes in which the costs and values of inputs and outputs vary more rapidly than the human operator can fully comprehend. His first responsibility is to maintain safe, dependable, high-performance operation. The further condition that the control be as profitable as possible is a less exacting, less well-defined requirement which can be met more effectively with a good knowledge of the cost equation relationships involved.

5.6 Cost-Benefit Analysis

In the preceding sections the matter of cost has been considered from the aspect of establishing the costs for a fairly well-defined system. Another large class of problems involves the selection of preferred system characteristics in situations where economic resources, particularly funds, are a significant constraint and where important technical or operational trade-offs may involve significantly different requirements for such resources. This type is exemplified by a number of cost studies that must be made in the space and military areas and has been given the name of cost-benefit analysis or cost-effectiveness analysis.[52] The basic objective in these studies is to elect, from among alternatives, the "optimum" or, to be more accurate, the "preferred" system to perform some given mission. The degree of performance or effectiveness of that mission is, of course, a central consideration and must receive careful attention. In addition, where economic resources, such as funds, are an important constraint, the need for these resources also becomes a basic criterion for selecting system characteristics.

It is a primary thesis of this presentation patterned after that of Harry Hatry that, to make a rational selection of system characteristics, cost and effectiveness criteria *both* need to be considered and to be considered jointly. Frequently, technical characteristics have been the only or at least the predominant consideration during the process of selecting system characteristics. Costs have been used primarily as an overall constraint rather than as a measure to assist in evaluating trade-offs.

The major steps involved in the cost-benefit analysis are as follows:

1. *Identify pertinent measures of effectiveness.* Measures of effectiveness must, in general, be quantifiable, realistic, and meaningful. Mission statements such as "to provide appropriate command and control over the country's airborne traffic" or "to provide adequate defense against hostile submarines" are clearly too vague to be of much use. Such statements have to be translated into specific, measurable terms. Some examples from studies are shown in Table 5.6-1.

Table 5.6-1. *Illustrations of Measure of Effectiveness*

Study Mission Objective	Measure of Effectiveness
System to repair orbiting satellites	Expected satellite downtime per year
Weather/reconnaissance satellites	Frequency of coverage of desired area
Tactical bombing aircraft	Number of targets destroyed Number of aircraft surviving
Nuclear detection system	Probability of detecting a clandestine detonation Expected number of false alarms per year

Note that frequently more than one measure may be needed. Sometimes it is sufficient to consider additional measures in a qualitative manner only. Utilizing more than one measure adds dimension to the results but may be quite justified.

With certain types of systems, the criterion problem may appear to be particularly nebulous. However, by probing the technical requirements and end products of the particular mission, satisfactory, though rough, measures can invariably be generated. For example, consider the problem associated in examining a lunar scientific exploration system. Initially it was felt that the matter was simply one of success or no success. However, it rapidly became clear that numerous alternative systems needed to be considered, each of which very probably involved different degrees of scientific payload, length of exploration time, safety, and probability of mission success, as well as different initial operational dates—any or all of which could be important indicators to be considered in system selection.

2. *Describe the alternative ways of performing the mission.* The alternative ways of performing the mission should be described in sufficient detail to permit satisfactory identification of the significant costs and of the factors affecting mission performance. The level of detail required is determined by the nature of the problem. For example, in studies to select preferred space systems whose operational date is not expected until ten years from now, it would clearly be unwise to attempt to define in great detail the hardware for the ground support equipment or the specific logistics supply system that would be needed, although frequently such details are requested. In contrast, if a logistics system was being selected for an operational submarine program, considerable detail as to the cost, size, and demand functions of the maintenance materiel would probably be needed.

3. *Distinguish the pertinent elements of cost.* Hardware costs are usually only one cost component, although likely to be significant. Most military systems involve many other expenses, including facilities, personnel pay and allowance, training, supplies, and various support equipment.

Equipment items require operating and maintenance expenses, as well as investment costs. Research and development costs for these items may also be involved.

Usually, in these studies, the costs that are pertinent are the "incremental" or, in the economist's term, the "marginal" costs.[23] That is, costs that have already been expended are generally considered to be irrelevant. A major effort should clearly be directed at distinguishing those cost elements that are most sensitive to the particular system characteristics to be chosen. In making a selection among alternatives, the difference in costs and effectiveness between the choices is of most importance. Nevertheless, since the approximate *total* cost of the preferred system is usually also of considerable interest and may affect the level of effectiveness that is obtainable within budgetary restrictions, even a relatively small difference in a major cost from one alternative to another, should not be neglected.

4. *Express mission performance and cost as functions of the primary system characteristics.* This step involves the development of the mathematical model which will reflect the significant relationships existing among cost, performance, and the system characteristics. For example, in space studies, booster cost is usually one of the significant cost elements; see Figure 5.6-1. The required booster thrust is some function of the payload weight. If specific boosters are not being

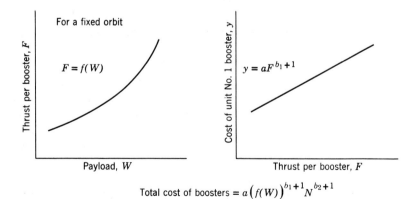

Total cost of boosters $= a\left(f(W)\right)^{b_1+1} N^{b_2+1}$

Figure 5.6-1. Booster cost equations.

Cost-Benefit Analysis

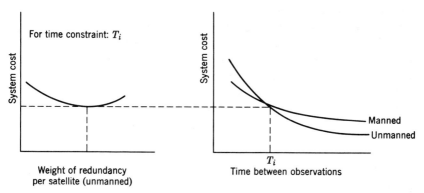

Figure 5.6-2. Suboptimization: global vehicle system.

used, the cost per booster of the No. 1 production unit can be expressed as a historically derived function of thrust. The total booster cost for the system can then be estimated by applying the appropriate learning curve[44] to the cost of booster No. 1 and the total quantity of boosters required. The particular values for the required thrust and the required number of boosters would probably be derived from other equations in the model, which would determine these requirements for specific levels of mission performance. The cost impact of various trade-offs (e.g., altering the size of the payload or the number of payloads) can be reflected through equations such as those shown on Figure 5.6-1.

If the alternative systems differ significantly, it may be necessary to develop separate sets of equations for each system. For example, in comparing a manned versus unmanned satellite system to perform a surveillance mission, separate equations can be prepared to reflect the unique operational concept of each alternative. At other times, the same equations may be appropriate to select the characteristics of the preferred system—it being necessary only to alter the values given to the parameters. For example, the same basic equation could be used to reflect the costs of a variety of boosters which could serve to launch a given payload.

Usually, in systems cost studies, both situations will occur. Generally, certain "suboptimizations" are required within each major alternative before these alternatives can be compared. For example (Figure 5.6-2), before comparing a manned and an unmanned satellite surveillance system, it was necessary to optimize within each system the amount of redundancy that each satellite would require. The amount of redundancy increases the useful lifetime per satellite, thereby reducing the number of replacement satellite launchings required. However, increasing redundancy increases the weight of each satellite, thereby tending to raise the cost per booster, the cost per satellite, and various other ground support costs per launching, thus, in general, increasing the total cost per launch. The graph on the left indicates the minimum cost point for the unmanned alternative for one specific level of effectiveness. This minimum point is then used to represent the cost of the unmanned system for that level of effectiveness. The computations indicated by the curve on the left were performed for various levels of effectiveness (that is, various times between observations), and the results used to plot the curves on the right.

In such suboptimization care should be taken to make certain that the effectiveness criteria used are consistent with the higher-level criteria.

Simpler cost models should be preferred to more complex models in cases where the marginal returns of the complex versions are at all questionable.

5. *Provide the specific values to be used in the estimating equations.* All cost analyses, directly or indirectly, utilize current and past experience, in other words, historical information. To the extent that the systems being evaluated are projected into the future or are of a novel nature, the usefulness of history is reduced. Extrapolations and qualitative judgments (inevitably based upon analogies to current systems) are needed to develop estimates of equation parameters. Statistical analysis helps, primarily in bringing order out of historical information, but has limited utility when data are applied to future system operations where the statistical "populations" are ill defined. For this reason, probably the essential ingredient for satisfactory system cost estimating is the presence of experienced engineering and cost analysis personnel who are able to relate known conditions to those of the hypothesized systems through their own judgment and with the aid of any mathematical tools that may be appropriate.

It should be noted that the parameters involved are not solely cost ones, such as "the salary per man per year." Also included are many physical and operational parameters, such as "the mean time between failure of a piece of equipment" or "the probability that a booster will be launched successfully."[62] Both types of parameters enter into the mathematical models required.

6. *Perform the necessary computations and analyze the results.* To illustrate the nature of the output of cost-benefit studies, consider

Cost-Benefit Analysis

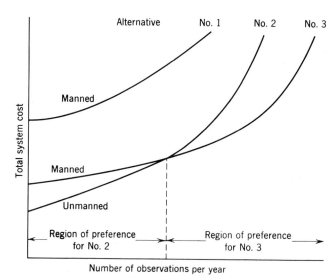

Figure 5.6-3. Cost-benefit analysis results: manned versus unmanned vehicle system.

the following example. The problem is a comparison of an unmanned versus a manned vehicle. Figure 5.6-3 shows for three alternative system designs a plot of cost against a single measure of effectiveness, the number of observations per year (the "reciprocal" of time between observations which was used in Figure 5.6-2).

Alternative No. 1 is a novel manned vehicle system requiring a high degree of initial investment, and therefore uneconomical for this mission within the range of effectiveness shown on the graph. Alternative No. 2 is the unmanned system, and No. 3 is another manned system, based upon modifying some of the hardware of the unmanned system and therefore not requiring as high an initial investment as alternative No. 1. The use of man in these systems permits certain economies in the maintenance and repair of the satellites, so that annual operating costs tend to be lower for the manned vehicles (as indicated by the slopes of the lines).

Considerable information is provided even on such a simple graph as Figure 5.6-3. First, the graph indicates for each effectiveness level the system which is least costly and by approximately how much. Second, it indicates for any given budgetary limitation what level of effectiveness can, and cannot, be achieved, and third, it indicates the cost penalties associated with choosing an alternative other than the least expensive system. For example, if alternative No. 1 was still a competitor for some other reason, such as an early operational date, the approximate cost or effectiveness penalties resulting from choosing this alternative would be shown.

It will be noted that each alternative is evaluated for various degrees of effectiveness. It is not always recognized that in most problems of system selection a number of levels of effectiveness are involved and should be considered since they may affect the decision. Seldom is it the case that only one fixed level of effectiveness need be associated with a mission.

The Problem of Uncertainty[62]

So far, little has been said about the reliability of the results of cost-benefit analyses. Clearly, this is a very important question about which, unfortunately, only unsatisfactory answers can be given.

Errors in the results can come from many sources, as shown in Table 5.6-2. The last source, computational and mathematical mistakes, probably can be avoided with enough care. The other sources of error, however, are probably impossible to eliminate since they essentially involve, at least in part, attempts to predict the future.

In the following will be discussed only the very important problem of the uncertainty in the values assigned to the parameters of the equations. Clearly, the further out into time for which the estimates have to be made (and in space studies, one is often projecting five to ten years into the future), the less confidence can be placed in these judgments.

The objective is not to eliminate uncertainty, which is impossible, but to quantify it. That is, the analysis should provide an insight into the effects of the uncertainties upon the decisions to be made.

One approach to this problem which has been applied in many of these studies is to employ sensitivity analysis. In this approach,

Table 5.6-2. *Sources of Error in Cost-Benefit Studies*

1. Incorrect or incomplete mission and measure of effectiveness determination.
2. Incomplete specification of the alternatives that should be considered to perform the mission. (The computational model, of course, can consider only the alternatives fed into it.)
3. Poor cost and performance functions and poor estimates for the parameters of the functions.
4. Incorrect estimates of the environment under which the system will eventually function (including the level of capability which will be required).
5. Computational mistakes in developing the model and in performing the calculations.

Cost-Benefit Analysis

the equation parameters believed to be subject to the greatest uncertainty and to be of the most significance to the study are given more than one value.

These values represent the range within which the actual value of the parameter is likely to fall. For each value selected, the calculations called for by the model are performed. The results can then be used to indicate the sensitivity of the system selection criteria to the value of the parameter. Sensitivity analysis thus represents evaluating the partial derivatives of the cost equations with respect to these selected parameters.[49] The computational problem rapidly becomes very great as more parameters are handled in this manner. This seriously restricts the amount of such sensitivity analysis that can be performed.

The results of this type of analysis are illustrated by Figure 5.6-4. This is a modification of the previous illustration, Figure 5.6-3, dealing with unmanned versus manned satellite systems. Bands are used rather than single lines to express the cost versus effectiveness relationship. The upper boundary of each band represents the circumstances which would result under the "most expensive" conditions; the lower boundary, the "least expensive" conditions. It is to be noted that, even with the ranges of uncertainties considered, alternative No. 1 is almost completely dominated by the other possibilities. Hence

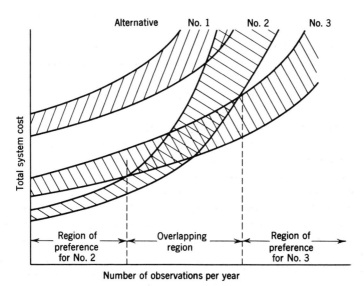

Figure 5.6-4. Sensitivity analysis of cost effectiveness.

the decision maker should feel a much higher degree of confidence that No. 1 really is a more costly alternative. It is also apparent that the decision maker's problem has become more complex; there now exists a large overlapping region between alternatives No. 2 and No. 3. In addition to merely considering specific levels of effectiveness and cost, it is now also necessary to take into account the likelihoods of the various conditions which cause the bands.

It should be emphasized that the *basic* problem has not been made more complex. We have merely expressed in a quantitative manner uncertainties that were always present. There should be little disagreement that the complexity of the problem reflected in Figure 5.6-4 is probably much closer to "reality" than the more simplified presentation of Figure 5.6-3.

The results should be examined to seek, wherever possible, the system alternatives that are economically preferable under a wide range of effectiveness and parametric conditions.

A limitation of this sensitivity analysis approach is that, although it indicates the *impact* of a change on the value of a parameter, it does not show the likelihood of that change. With sensitivity analysis, qualitative statements, at the very least, should be made about the relative likelihoods.

Another approach involves a more strongly statistical treatment. In this approach, the systems analysis provides a range of values for any parameters whose actual values are subject to uncertainty. These values for each such parameter are assumed to conform to a particular probability distribution, chosen with as much information as is available. By properly combining the resulting expected values and variances of each parameter in a manner consistent with the equations of the cost model, an expected value and variance for the *total* system cost can be estimated for each alternative.

As to the overall validity of cost-benefits studies, considering all sources of error, it appears that validation of their results is very difficult, if not impossible. The long-time horizons and changing system specifications (beyond the ground rules of the original study) have prevented other than crude attempts at verification. In any case, only the system alternative that is actually put into operation can be directly compared with the study results, since the other alternatives probably never come into existence. Despite the lack of rigor in arriving at such cost-benefit analyses, they do provide an orderly discipline for estimating costs which can be improved with continued experience. The methods previously employed have frequently resulted in greatly underestimated costs.

Conclusions

Because of the importance of system effectiveness to the military, an advisory committee was established to investigate this subject. The Weapon System Effectiveness Advisory Committee (WESIAC) has been engaged in the task of providing "technical guidance and assistance to the Air Force Systems Command in the development of a technique to appraise management of current and predicted weapon system effectiveness at all phases of weapon system life." Among this committee's results, recently published in a series of final reports,[50, 51, 52] are discussions of techniques for cost-effectiveness analysis.

The general conclusions reported by this committee include the following:

a. "Cost-effectiveness as an art, science, or discipline is still in its infancy.

b. "No single 'cookbook' method for performing cost effectiveness studies is possible at this time [1965].

c. "Task Group IV has shown that a general framework for performing cost-effectiveness studies on a more common basis, leading to standardization of methods and evaluation of results, is possible and very desirable.

d. "Standardization and availability of data to perform effective cost-effectiveness studies are seriously lacking.

e. "Many persons have little knowledge or appreciation for not only the usefulness but also the intent of cost-effectiveness studies.

f. "More study and education of the methodology and techniques for performing cost-effectiveness studies are needed."

The terms cost effectiveness, cost benefit, and value effectiveness tend to be related to the common problem of endeavoring to compare quantitatively the value of the results from a system and the cost of it. The material in this section has served to illustrate an overall approach to cost-benefit analysis and to emphasize the importance of being able to do the cost-effectiveness job better in the future.

5.7 Conclusions

Cost can be a very critical factor in the evaluation or judgment of a system, and therefore its determination should and does receive a great deal of attention. Furthermore, since costs are more easily understood than systems performance or reliability by the persons concerned with the purchasing, financial, and business affairs of the system, who frequently have a major influence on its purchase, it is important that the overall functional implications of the costs over the life of the operating and support systems be clearly demonstrated.

There is a wide disparity between the costs involved in mere study of a system in a preliminary fashion and those required for the research and design of a system which is to go into production. Clearly the costing process must reflect the cost for the product or service in question, and that in turn must be related to the extent of the operational and/or support function being performed. The possible influence of other concurrent jobs or systems, as well as alternative or competing systems and methods, on costs should also be taken into account in the costing process. The effect on the cost of the system of decisions covering the time range from seconds to years should likewise receive proper attention.

To obtain proper values it is beneficial to build up the total cost from the cost of the various parts, for example, costs associated with initial system investment, other fixed costs, variable costs, and past investments and other charges. In a number of cases where fixed costs are essentially independent of the medium- or short-term cost considerations, decisions regarding these medium- and short-term costs can be more easily and speedily made with a minimum regard for long-time or fixed-cost values.

An area of increasing importance in system cost estimating is that relative to cost-benefits or cost-effectiveness. The need for creating suitable, quantitative, effectiveness measures has been emphasized, and illustrations have been provided to demonstrate the nature of the results to be obtained for both deterministic and probabilistic input data.

6

Time

6.0 Introduction

Time is an important "fourth" dimension in systems engineering. As pointed out in Chapter 1, changes are continually taking place with time. It is important that the timing relationships between the different parts of a system be appropriate, and also that the time involved in engineering and making the system itself be predictable and proper.[35] The advantages both technically and commercially of being first in time are well known; the penalties of being late are also clearly understood by all. It is of the greatest importance in systems engineering that a thorough understanding be achieved of how time affects the system.

The subject of time is one that has received broad philosophical treatment, and it is of interest to note a few of these general observations in this introductory section. Omar the Tentmaker wrote, "The Moving Finger writes; and, having writ, moves on." In keeping with the concept of time, this serves to emphasize the constancy of change and the need for continual reappraisal of future plans based on the current evaluation of past and present facts. The best of plans are only plans; as the future unfolds to become the past, we must adapt new plans for the future based on the feedback of recent information from what has already happened.

Arnold Bennett mentions the oft-stated wish, "If I only had more time." He points out, "We shall never have any more time. We have, and we always have had, all the time there is," 24 hours a day. Traditionally, we think of 8 hours of these 24 during 5 of the 7 days a week as being available for work. Actually, that accounts for only 40 of the 168 hours a week. There is more time than most of us use, but there is only a finite amount of it.

Two sayings pertinent to the subject of time are "We get old too soon, and smart too late," and "Do it now!" Both of these serve to point out the necessity for timely action. Although some problems solve themselves if not acted upon, many remain sources of difficulty and require attention even if we refuse to take action. How many times it is found that a problem will take only a few hours of attention to resolve, even though it may have been untouched for weeks or months.

Last is the thought that "the human time constant is the longest one in any system." Frequently it takes the longest time to get people to make up or to change their minds. Once this is accomplished, progress can be made. Because of this inherent time delay caused by people, an effective communication means is essential to ensure that as many persons as possible who are concerned with a system understand the overall plan and the current status of the different parts of the plan.

This chapter will explore first a number of the different meanings of the word time as used in a system sense. Then the general effect of using more or less time on the other major system factors of performance, cost, and reliability will be discussed. The next topic will be time as considered from the two-boundary-conditions point of view; namely, how one can accomplish a given task in the minimum time from the present, and how one can accomplish a task for which the only criterion is that it be completed by a certain date—i.e., the terminal value problem.

Attention will be given to the time required to make the system as well as to the time for the system to perform. Time schedules and the importance of their monitoring will be discussed. A presentation of PERT, the popular Program Evaluation and Review Technique, will be given. Also included are a description of how PERT works and how PERT-like consideration is being given to PERT-COST.

6.1 Different Meanings of Time

As in many semantic problems, the word time is used by different people to mean different things. Such concepts as total elapsed time, time required to perform the actual task, minimum time, and many others are frequently referred to merely as "time." Hence arises the possibility for confusion. Reference to Figure 6.1-1 will help to illustrate the time-ambiguity problem.

The *development time* required to develop and design the system is devoted to the engineering and preparation of the necessary manufacturing and other instructions. The *production time* is necessary for preparation of the facilities as well as for actual manufacture

Different Meanings of Time

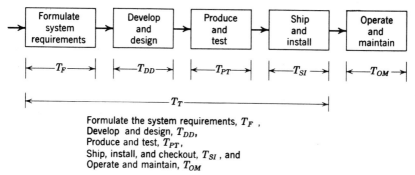

Formulate the system requirements, T_F,
Develop and design, T_{DD},
Produce and test, T_{PT},
Ship, install, and checkout, T_{SI}, and
Operate and maintain, T_{OM}

Figure 6.1-1. The time associated with different functions involved in making, installing, and operating a system.

and test. The *ship, installation, and checkout time* is needed to accomplish the shipment as well as the preparation of the facilities for installation and checkout.

It is apparent that the time required for the preparation of facilities may be much greater than the time required for their use. Furthermore, the shorter the operating time cycle desired for any of these parts, the longer may be the preparation time necessary to get the facilities into place. We have a systems problem here of striking a proper time and cost balance between how much time and money can be justified for preparation and how much for use of the various systems.

Formulation time represents the time required by the system-user as well as the system-maker to conceive and determine just what characteristics the system should have.[43] Frequently this period also involves trying to answer such allied questions as the following: Is this system necessary? Can it be afforded and, if so, at what level of cost and performance? Which of several willing and eager systems makers should be selected to do the job? As a matter of interest, the task of formulating the system requirements may turn out to be one of the more time-consuming portions of the overall job required to get the system into being and into operation.

The *operation and maintenance time* for the system will differ widely from a major industrial system that may last 5–20 years to a military system which may have a much shorter period of operation but a long time of ready-maintenance.

As noted in Chapter 5 on costs, the matter of whether the end stage of a particular system is to result in feasibility, prototype, or production equipment may have a significant effect on the time required for the different parts of the systems job described in Figure 6.1-1. It should be evident that, if a significantly different (larger or smaller) cost is associated with a particular task in comparison with the other ones, there is a good likelihood that the times required will also be different (larger or smaller) though not necessarily proportional.

The various times, T_F, T_{DD}, etc., indicated in Figure 6.1-1 are assumed to be *elapsed time*, or calendar time, associated with each part of the system effort. These times represent the number of days, weeks, or months that elapse between the start of each particular part and its completion. Elapsed time is significant from the point of view of scheduling one part relative to another.

Not all of the elapsed time is gainfully occupied in accomplishing the task involved. The times t_F, t_{DD}, t_{PT}, t_{SI}, and t_{OM} are used to identify the *time to perform the actual work associated with each task*, where t_F/T_F, t_{DD}/T_{DD}, etc., are ≤ 1.0. Because the whole 24 hours per day are not used and because time is required for queuing and time sharing of such critical resources as management, men, and machines, the ratio of t/T can frequently be very much less than 1.0. From a resources management and production planning point of view it is essential that the times required to perform the actual work (t's) be known and differentiated from the elapsed times (T's).

The *total elapsed time* to get the system from its inception at time of formulation to its readiness to operate after installation and check-out is shown in Figure 6.1-1 to be T_T. In general this total elapsed time is less than the sum of the elapsed times of the various parts: formulation, development, production, and installation and check-out,

$$T_T < T_F + T_{DD} + T_{PT} + T_{SI} \tag{6.1-1}$$

T_T can be less than the sum of its parts because there can be and generally is a certain degree of overlap of time in which some of the individual major tasks are done concurrently (in parallel) rather than being handled in a consecutive (in series) fashion. Figure 6.1-2 illustrates the overlapping of time before the completion of one task and after the initiation of the next phase. Although from the point of view of shortening the total elapsed time, T_T, this overlapping appears desirable, there is a certain amount of risk that the work done on the following task will be incorrect or wasted because full and correct information may not be available at the time of initiation of the following task. However, by judicious selection of the parts of the succeeding tasks to be initiated first, starting the later tasks at an earlier time can frequently be used to good advantage to de-

Different Meanings of Time 225

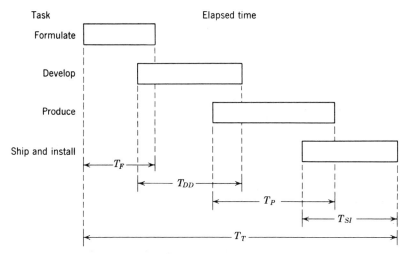

Figure 6.1-2. Total elapsed time to obtain a system, shown to be less than the sum of the times to do its parts.

crease the total elapsed time without too adversely affecting the end result in performance or cost.

In the preceding consideration of elapsed time, the emphasis has been on the time required to perform the system design, fabrication, and installation. From the user's point of view, often the *operating time cycle* of the system, which forms a part of T_{OM} in Figure 6.1-1, is the most important time involved. By operating time cycle, one refers to the time which it takes for the system to perform its task. For example, in an airlines reservation system, how much time in minutes or seconds does it take to make a reservation, or conversely, how many reservations can be handled per hour; in a warehousing system, how much time in minutes or hours does it take to fill an order, or conversely, how many orders can be filled in an hour? Although the elapsed time required to design, produce, and install the system in the first place may be different for systems having different operating time cycles, a reduction in the operating time cycle may be far greater than the proportional increase in elapsed time to obtain the system in the first place. For example, putting in a bigger motor may make it possible to shorten the operating cycle, but the increase in time required to produce, ship, and install the modified system may not be proportional to the change in motor size.

Two other important expressions incorporating the word time are *minimum time* and *terminal time*. Minimum time, T_{MIN}, describes

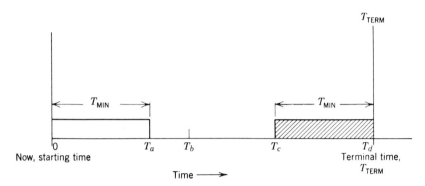

Figure 6.1-3. Sketch describing several time options available between now and the terminal time.

the shortest value of elapsed time from the start of a task to its completion. Terminal time, T_{TERM}, indicates the particular value of clock or calendar time when a specific event or task should occur or have been completed. The term *slack* (or slack time), used as a measure of time to spare, often appears in connection with terminal time where some of the time between the starting time and the terminal time is not gainfully utilized, that is, it is slack time.

Figure 6.1-3 illustrates in a simple fashion some of these time expressions. Now, the starting time, is shown to be zero. Some time in the future is the terminal time, T_{TERM}, when it is desired that the task be completed. The minimum time, T_{MIN}, is shown to be an elapsed time of T_a units. Thus, if the task is started now, it will be done at T_a units of elapsed time. If, on the other hand, work is not started until T_c units of elapsed time (that is, there are T_c units of slack time between now and the start of the effort because of assuming the risk that the job will take the minimum time), it is still possible to have the task accomplished by the desired terminal time, $T_d = T_{\text{TERM}}$.

Of course, still another intermediate condition is possible in this case, where $T_{\text{TERM}} > T_{\text{MIN}}$. This is to start the task at elapsed time T_b, i.e., after T_b units of slack, perform the task in some time less than $T_d - T_b$, and have still more slack after completion of the task and before the desired terminal time, T_d.

These simple illustrations serve to demonstrate the fact that there may be alternative time solutions as well as alternative performance or cost solutions for a systems engineering problem or part of a problem. The decision is frequently up to the system engineer to choose the sets of possible conditions from a time sense that are acceptable and favorable to the purposes of the system or undertaking.

The many different meanings of the word *time* just discussed serve to emphasize the various facets of the system's sensitivity to the whole concept of time. The term has many different interpretations in the frame of reference of the different individuals and groups involved. Indeed, since all parties concerned share this time in common, it is important to keep in mind the interaction or interference that may be taking place. It is in many cases as real as if it were a mechanical or electrical interference. The system engineer has an important responsibility to control the various times as well as the other factors, such as money and performance, involved.

6.2 Effect of Time on Other Factors

The time objective selected to accomplish a project may influence the nature of the other factors that characterize the system—its performance, cost, and reliability. Although the following ideas are largely qualitative in nature, they serve to indicate some of the considerations that influence these other factors as the goal set for the total time to complete a project is changed from the generally accepted "nominal" value.

To place in proper perspective what happens to the resultant system when the desired total time to complete the project is changed, let us first consider the effect on performance and cost as a function of elapsed time for the nominal case. The solid line on Figure 6.2-1 shows for such a case how a production system increases in actual performance capability as a function of time. For the first 60–70 per cent of the nominal time, effort is going into preparations for producing the system, so that the actual performance capable of being obtained by production equipment is not significant. Only as the nominal time for completion of the project approaches will the production system begin to meet all its performance requirements. Furthermore, once these requirements are met, the amount of performance capability for a well-designed system tends to level off, since the system characteristics have been engineered to meet the nominal performance characteristics. For example, a system designed to handle a load of so many units per hour may be able to accommodate 10 or 20 or even 50 per cent more units per hour, but the system which has been "optimized" generally can only accommodate so much more overload.

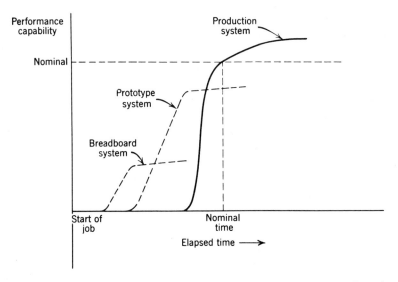

Figure 6.2-1. Performance possible from a breadboard, prototype, and production system as a function of elapsed time.

Also shown by dotted lines on Figure 6.2-1 are the performance capabilities of the breadboard and prototype systems. The breadboard may be available in one-quarter or one-third of the nominal time, but it generally displays only a small part of the total nominal performance if environmental and other allied characteristics are included. The prototype system may be available in about two-thirds of the nominal time and more nearly approaches the performance capability specified. Since it has probably not been made in the same fashion or with all of the same equipment as will be the production equipment, its capability may not be the same. Also, experience in operating and testing the prototype will doubtless reveal some deficiencies which should be removed in the production system.

Shown on Figure 6.2-2 are curves of the rate of cost expenditures, as well as the cumulative cost, as a function of elapsed time expressed in terms of nominal. These curves show an initial delay in costs in starting up the effort; the major effort tends to take place at a high and increasing rate, and the shut-down period tends to be a tapering-off one extending beyond the nominal completion time.

From Figures 6.2-1 and 6.2-2, one notes that there is a lower time limit below which it is impossible to accomplish the performance objective without appreciably increasing the costs. Also, the expenditure rates just before the nominal completion time are high. Hence,

Effect of Time on Other Factors

if one misses the nominal time estimate, costs are being incurred at a high rate, and the total cumulative cost will tend to rise rapidly if the system is not completed approximately on schedule.

Let us now consider what may happen to the performance, cost, and also reliability if the nominal total time to accomplish the system project is chosen initially to be either shorter or longer than for the normal case considered above. The situation being considered may be stated as follows. A new system is proposed, and the normal time estimated to obtain it with a nominal level of performance, cost, and reliability is 10 months. What would be the comparable performance, cost, and reliability if the decision were made to achieve the same system in from one-half to two times the original nominal time, i.e., 5 to 20 months?

Of course the degree of motivation of the people involved, their sense of challenge, and their previous experience and knowledge of the type of system called for will all influence the results to be obtained. However, the following represent some reasonable estimates of what might be expected.

Performance

Figure 6.2-3 shows the performance versus total time to complete the project and indicates that for the range of one-half to two times

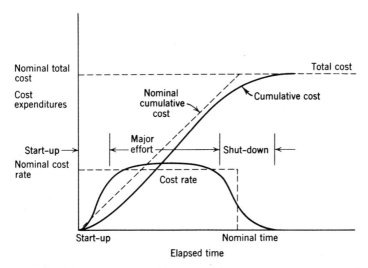

Figure 6.2-2. Rate of cost expenditures and cumulative cost as a function of time from start-up.

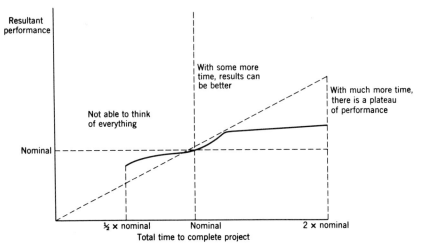

Figure 6.2-3. Resultant system performance as a function of total time to complete project, compared to nominal.

the nominal time a relative gain can be realized by decreasing the total nominal time. That is, the performance tends to approach the nominal level, but the time required is less.

For the shorter than normal values of nominal time, the performance tends to approach the original nominal value because of the inherent requirements and capabilities established for the system. However, with the shorter amount of time it is not possible to think of or take care of everything, and a somewhat lower performance results. People working sustained periods of overtime do not perform at their best efficiency and are not able to be as creative as if they are allowed to work at a more normal pace.

Since the basic capability of the system has been established by its requirements or specifications, there is an upper limit or plateau on how much performance can be obtained with considerably more time, say twice the nominal. However, with a little more time, say up to 30–50 per cent, modest improvements in performance can be made when there is more time to do a more thorough design, fabrication, test, and installation job.

Cost

The corresponding picture of cost versus total time to complete the project is given in Figure 6.2-4 and shows the cost to increase for total times appreciably shorter or longer than the nominal. When the nominal time is cut in half, roughly the same amount of work

Effect of Time on Other Factors

is still required, so overtime work or additional people or both are required. Overtime costs are at higher rates than normal, and new people are generally not able to contribute as effectively as the ones previously familiar with the work. Because of the pressure of getting the job done appreciably sooner than normal, there will be a tendency for more mistakes, which in turn will further increase the costs. The overall effect will be to raise the costs well above those for the nominal case.

Interestingly enough, for a minor decrease from the nominal time, the total cost may be slightly less. Although some overtime costs are incurred, these are more than offset by the increased effort that is early recognized as essential to get the job done on time. The job becomes a "priority" one and may get preferential service to meet a schedule that is unusual but capable of attainment.

For times somewhat greater than nominal, the cost will tend to increase proportionally to the increased time because work has continued to achieve the increased performance noted in Figure 6.2-3. As the nominal time is extended to approximately twice the normal value, the costs as well as the performance tend to level out, but the costs are at a higher level than the increase in performance. Employees not working full time on the job tend to charge for their

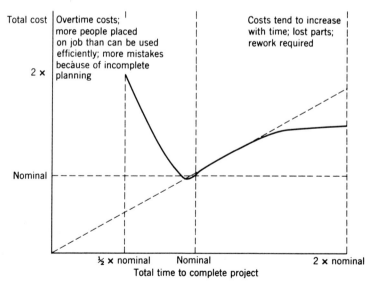

Figure 6.2-4. Total cost as a function of total time to complete project, compared to nominal.

time as though they were, parts get lost and have to be remade, and more rework is required because people tend to spoil more parts as a result of forgetting where they were before they were taken off to do something else.

Reliability

Although system reliability is discussed in much more detail in Chapter 7, it is of interest here to consider briefly how this factor is affected as a function of the time to make the system. The reliability versus total time to complete the project is estimated to be as shown on Figure 6.2-5 and is indicated to be strongly proportional to the total time for completion. Because reliability is so dependent on details and careful attention, when the time cycle to accomplish the work is shorter, the tendency for a less reliable system is greater. Conversely, with more time it is possible to do a more thorough job of design reviews, additional laboratory testing, etc. Of course, at a certain point there is again a leveling off, which might in fact begin to decrease if the total time were increased still further.

To see the reliability picture in another way, consider the experimental data shown in Figure 6.2-6, which are the results from experience in developing a new line of d-c generators.[44] As indicated, ap-

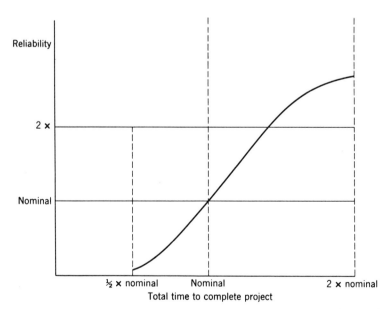

Figure 6.2-5. System reliability as a function of total time to complete project, compared to nominal.

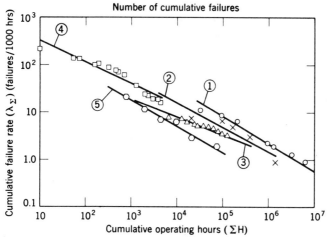

Figure 6.2-6. Cumulative failure rate trends as a function of cumulative operating hours. J. T. Duane, Learning Curve Approach to Reliability Monitoring, *IEEE Trans. Aerospace*, **2**, No. 2, April 1964.

preciable gains in reliability can be realized as a function of increased work with a given design of equipment, since reliability is so closely concerned with the elimination of small and difficult-to-locate mistakes.

The curves on Figures 6.2-3, 6.2-4, and 6.2-5 are merely estimates of how the performance, cost, and reliability might change with altered values of the nominal time. They assume that the basic system to be built in each instance is essentially the same. They do not allow for a radically different way of achieving the system, which might result from a realization that something unusual is required to meet a shorter time schedule or that some innovation is possible with a greatly extended time schedule. Although these types of curves will differ for different systems, it is worthwhile to try to consider what their shapes might be for each particular system being planned.

6.3 The Two-Time Boundaries—Now or Then

The question to which this section addresses itself is that of the best time to do something—whether a particular action should be done *now* or at some later time, *then*. Since most systems problems involve many variables which at any particular time have varying degrees of time criticality, it is important to determine the sensitivity

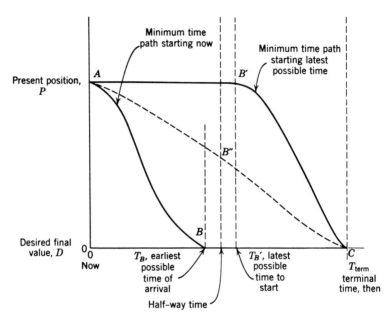

Figure 6.3-1. Comparison of alternative ways of going from present position to desired final value by the terminal time, T term.

of each specific variable with respect to time. In the sense of linear programing, the two limits in time are the first possible chance—now, and the last possible chance—when this job is absolutely needed. Although in the sense of compromise the truth (and most desirable way) may lie somewhere in between, let us consider what happens in the two extremes.

The problem being considered is illustrated in Figure 6.3-1. At the present time (now), $T = 0$, the present condition of our activity is represented by present position, P, shown as A on Figure 6.3-1. On or before terminal time T_{term}, the position of our activity is to be at the desired final value, D, shown as point C.

It is assumed that the capability of accomplishing the activity is limited to certain finite maximum values of acceleration, velocity, and deceleration. That is, the maximum amount of energy or other resources available to accomplish the task is assumed to be fixed at a certain value. It is also assumed that the desired final value of position, D, and the final value of time, T_{term}, are known and fixed. Furthermore, it is assumed that there are no external disturbances or other influences which cause the activity to occur in a

The Two-Time Boundaries—Now or Then

manner different from the way presently planned. Later, the effects of these assumptions will be given further attention.

If, at the present, the full resources available are applied to accomplish the task, the position-time path is from A to B to C. The resources are used at their full capacity from 0 to T_B, and are not used at all during the "slack" period from T_B to T_{term}, the terminal time when the activity is really needed to be completed.

If, on the other hand, nothing is done from now until $T_{B'}$, the latest possible time to start, and then the full resources are applied from $T_{B'}$ to the terminal time T_{term}, the position-time path is from A to B' to C. In this case the slack period is of the same duration as in the previous instance, but it is taken at the start.

Whereas in the case of doing it *now* the time to start was decided upon by definition, in the case of doing it at the last possible time, *then*, the time to start, $T_{B'}$, is itself a factor to be determined. The determination of the time $T_{B'}$ is most easily arrived at by referencing time and position from the terminal time and desired final position and calculating backwards in time to where one could be, with the resources available, and still reach the desired final value on schedule. Figure 6.3-2 illustrates the results of this series of calculations.

The solid line passing through C shows the position-time path by

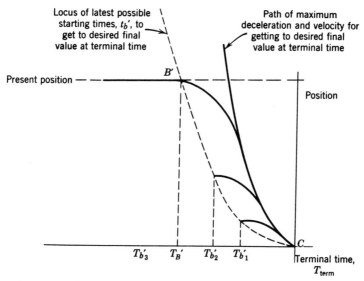

Figure 6.3-2. Loci of various paths of maximum deceleration and velocity for getting to desired final value from different initial positions.

which it is possible to reach the desired final value at the terminal time if one uses the maximum deceleration from speeds not in excess of the maximum speed. The dotted line shows the locus of latest possible starting times, $t_{b'}$, to get from these positions to the solid-line path. The time $T_{B'}$ is determined by the intersection at which the present-position line crosses the dotted locus of starting times. It is of interest to note that not only must $T_{B'}$, the time to start the activity, be determined, but also some later time at which the activity must be decelerated so that it may arrive at the desired final value at the terminal time specified.

Also of interest in Figure 6.3-2 is the fact that the locus of latest possible starting times is in effect the location where the maximum effort in the sense of Pontryagin's maximum principle is required.[49] That is, given a requirement and a certain maximum effort that is available, the locus of points for which this maximum effort is adequate is the dotted-line locus.

Referring to Figure 6.3-1, in addition to the two minimum time paths ABC and $AB'C$, there are other paths, such as $AB''C$, which represent still other allocations of resources and time which will also accomplish the same activity of getting from A to C with the same terminal time, T_{term}, but which use less peak resources at any particular time. Figure 6.3-3 shows a plot of peak resources required for different paths starting "now" at present position, P, at A to get the desired final value D at time T_{term} at "then." If the minimum time of actual movement is used, then the straight solid line shows that for path ABC or $AB'C$ or another path from some intermediate starting time the peak resources required will be the same. However, if the minimum peak resources are used, the dotted curve results, and up to starting time $T_{B'}$ the peak resources curve is lower than for minimum time to change. For still later starting times, Figure 6.3-3, the only way in which the desired activity or motion can take place is to raise the peak resources over and above those originally allocated. From the point of view that stresses how sensitive the result is to changes in the peak resources which are available, it is apparent that by starting "now" one can be most tolerant of changes in the peak resources. If one waits until $T_{B'}$ or later, assuredly only a little tolerance is available in the peak resources necessary if the estimated time T_{term} is found to be less than that originally determined, or if the measurement of present position is found later to have been greater than was originally thought to be the case.

As has been shown in Chapter 2 of *Systems Engineering Tools*,[49]

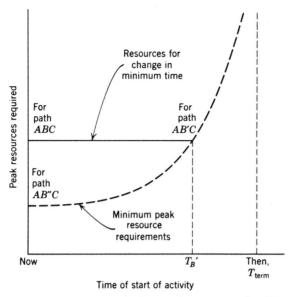

Figure 6.3-3. Peak resources required to go from A to C for different times of start of activity.

it is of the utmost importance that sufficient power (resources) and time be available within which to accomplish the desired action. It is essential that the basic capability of the resources be such that the desired action can meet the time and performance requirements.

The process of locating $T_{B'}$, the last possible time for switching on Figure 6.3-1, may itself be quite a time control, employing predictive control or dynamic optimization methods to establish when $T_{B'}$ should occur and also when the acceleration should be *reversed* to bring the equipment to its final terminal position. For the present purposes, the fact of importance is that stringent means may be required to achieve the control of minimum time that is desired.

6.4 Time to Make System versus Time Required for System Itself to Operate

Another area of significance where a possible confusion may exist in connection with the use of the word time involves the differentiation between the time required to engineer and make the system and the time required for the system itself to operate or to perform its function. Referring to Figure 6.1-1, one notes that the time required to engineer the system is composed of the first four blocks, namely, formulate the system requirements, develop and design, produce and test, and ship and install. The operate (and maintain) time shown by the block to the right may indicate the actual time of a single operation of the system (a matter of a few seconds), the time of operation of the system between maintenance (a matter of weeks), or the total useful lifetime of the system (a matter of years). Although it is obviously not possible to place a single value on each of these time factors for every system, it is useful to obtain an order-of-magnitude estimate for some of these values for some systems.

Time to Make System

Before proceeding to place estimates on the time to make a system it is interesting to refer to the book *Systems Engineering* by Goode and Machol, where the job of systems engineering is divided into two major parts, exterior design and interior design, as shown in Figures 6.4-1 and 6.4-2. Figure 6.4-1 shows the exterior design to be the statement of the problem, while the interior design is the suggested solution. In more detail in Figure 6.4-2 are shown a number of the individual tasks which comprise both the exterior and interior portions of the activities which must be done to make the system. In terms of the activities indicated on Figures 6.1-1 and 6.4-2, there is a rough correlation between the formulation process and the exterior design problem. Although from Figure 6.1-1 the develop and design, produce and test, and ship and install activities complete the time to make the system, the interior design problem of Figure 6.4-2 really covers most nearly only the develop and design portion of this remaining time to complete the system. However, interior design is said by Goode and Machol to include prototype construction as well as test, training, and evaluation phases, so that from the point of view of elapsed time it is apparent that interior design extends into the installation period. Obviously the choice of activities used here to describe the time to make a system differs from that used in *Systems Engineering*.[16]

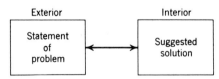

Figure 6.4-1. Interaction of exterior and interior aspects of design.

Time to Make System versus Time Required for System Itself to Operate 239

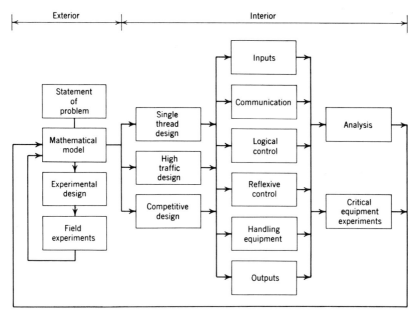

Figure 6.4-2. More detailed description of activities included in exterior and interior aspects of system design, from Goode and Machol.

Time to Formulate

The time to formulate may differ appreciably between various systems and will be quite dependent on the magnitude of the cost and performance expected. Although formulation may require as little time as a month or two, more frequently the job of understanding the problem and preparing specifications takes as long as one to two years. Since large sums of money or large numbers of persons have not yet been committed, there is a tendency for this time to drag on as people "try to make sure" that their specifications are complete. Meanwhile, as time passes by, some of the original advantages in terms of a novel solution present at the start of the formulation are being dissipated as competitors consider embarking on similar projects. Also, since the customer is not familiar with the design and manufacture process, he is somewhat handicapped in performing the formulation, but he tends to be reluctant to pay a vendor to help him in this task.

Time to Develop and Design

The processes involved in designing and developing the system, described in Chapter 2, can be long and painstaking for a completely new, large system. For systems which are merely modifications of existing ones to meet somewhat different customer applications, the time scale can be considerably shorter. The time for design and development may range from as little as 6 months to as much as 5–10 years. This period is one of high engineering manpower and cost requirements.

Time to Produce and Test

With the advent of computers, programed machine tools, and automated assembly and test methods, the time to produce and test the system, once it is decided what the system is to consist of, may not be the longest time involved in making a system. Although material costs are high during this period and different groups of people, namely, production and shop personnel, must be brought into the operation, time cycles of the order of 6 months to 2 years are reasonable for producing and testing the equipment and perhaps the system itself.

Time to Ship and Install

Depending on whether the system is small enough or is otherwise capable of being tested completely at the factory before shipment and installation, or whether it is extremely large or is for other reasons required to be assembled on the site, the time needed for shipment and installation may vary appreciably. To the extent that the initial design as well as the factory testing has permitted debugging the system before shipment, the shipment and installation time may be as short as 3 months. However, construction hold-ups as well as on-site engineering and "tune-up" may lengthen the shipping and installation time to as long as a year. In any event, installation may take a significant period of time, and proper attention should be given to installation problems during the initial system engineering phases of structuring and design.

Obviously, the times mentioned above are merely indicative of the range that may be required. However, they serve to point out that the total elapsed time to get a system to the point where it is operational is generally a matter of years, and cutting corners on the initial phases may result in a longer overall time to achieve a system that works satisfactorily.

Time Required for System Itself to Operate

The speed of the system or the time required for it to complete an operating cycle may be more a function of the performance speci-

Time Schedules

fications or requirements than of the time to make the system. This, of course, assumes that the system is within the state of the art so that one need not wait for a new invention or discovery to accomplish the desired results.

As described in Chapter 2 of *Systems Engineering Tools*,[49] as well as in Sections 6.2 and 6.3 of this book, the performance capability obtainable from a system is dependent on such factors as its power capabilities, its materials limitations, and/or its information or control rates. Thus, the original cost, accuracy, space, weight, and general state-of-the-art considerations determine to a large extent the time required for the system to operate, whether it be hours, minutes, seconds, milliseconds, or even shorter periods of time. In general, there does not appear to be a strong relationship between the time to make the system and the time required for the system to operate.

Another time of significance in connection with operation is the life duration of the system. Here reliability and obsolescence are major influencing factors. In the next chapter the reliability and maintenance effects on system life will be mentioned. Although for some depreciation purposes the life for write-off of system equipment may be as short as 2–3 years, generally longer use is contemplated, and system life of 5–20 years is frequently considered as reasonable for practical operational purposes.

6.5 Time Schedules

Time schedules are frequently built around the implied time structure that has been incorporated in the basic organization of the sys-

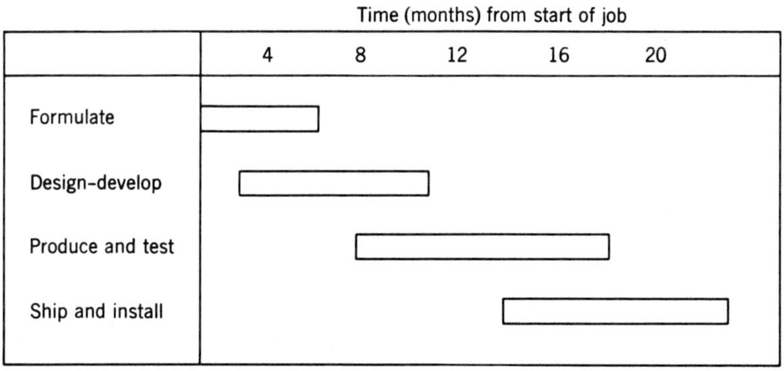

Figure 6.5-1. System time schedule chart in simplified form.

tem. This organization will have described different phases for the job, such as the formulate-design-produce-and-install cycle mentioned earlier in this chapter. Breadboard, prototype, and production constitute another structural organization around which a time schedule might be drawn.

The time schedules are generally shown in chart form as illustrated in Figure 6.5-1. The different phases of the work are listed, and the period and duration of the time required to do each phase are shown by the location of the appropriate bar. In general, as was noted in Chapter 2, it is necessary and possible to have the work on different phases take place with some overlap in time. Frequently from the specifications and from the general planning that takes place early in the system engineering process the general requirements of the later efforts are sufficiently well known for work to be started without having the final details that may depend on completion of the earlier steps. When enough information is available to go ahead, a reduction in the total elapsed time may be realized if an overlap of the time schedule is programed.

The time schedule of Figure 6.5-1 is a coarse one showing the overall system schedule. For greater effectiveness in planning and monitoring the detailed execution of the systems effort, it is necessary to prepare a more detailed schedule of each of the major tasks that make up each of the phases.

For a large system, an important psychological and practical factor is the assignment of the job of schedule preparation to each of the managers or responsible project leaders that are directing the various phases or major tasks. This schedule preparation should also show manpower and other supporting assumptions relating to critical resources. When the individuals involved have had a hand in the original schedule preparation, their involvement in the job and sense of responsibility for holding to the schedule are greater, and there is more likelihood of meeting the target dates.

Of course the final schedule will have to reconcile the overall system requirements, and therefore it may have to be changed from the first approximations of the individual managers or project leaders. This reconciliation and modification of individual schedules is the responsibility of the overall system manager or project engineer. In any event a composite schedule of the overall project and the significant events that compose it is highly essential to the realization of a satisfactory method of predicting, monitoring, and controlling the time of accomplishing a systems project. By the use of periodic checks in time, as well as at significant stages of system accomplishment, it is possible to provide a systematic feedback method for closing the loop between the scheduled time plans and the actual performance on the job of the different stages of designing and making the system.

In a way a schedule is a projection ahead in time of the planned course of the project and as such is like a model in a predictive control scheme. It provides a basis for judging the future results of present action and therefore allows the person or group responsible for the success for the venture to be aware of the effect of current events on the ultimate objective of the system. It is important to realize that schedules are only a plan for future action and that they are most valid as long as they reflect what is happening. If subsequent events indicate that there has been a slippage or falling behind of a schedule, the person responsible for the project should try to get it back on schedule by extra effort in the form of overtime work, added manpower, or other appropriate means. If these measures are not successful, it is advisable to revise the schedule in a conspicuous manner so that all parties involved are conscious of the new time plan.

6.6 PERT[33, 49]

In recent years the use of schedules for the management of the time for accomplishing a systems project has been complemented by the addition of the Program Evaluation and Review Technique (PERT) and/or methods for Task Network Scheduling (TANES). These methods expand the detail of schedules and provide other useful information to help the systems manager control the time that it takes to realize the project. PERT[14] was developed in the 1956-57 period by the management consulting firm of Booz, Allen and Hamilton for the U.S. Navy's Special Project Office to assist it in accomplishing the extremely complex production requirements of the *Polaris* nuclear-powered submarine project on schedule and within the cost budget. TANES, which will be discussed in the next section, was developed as a generalization of the techniques for analyzing activity and event networks such as are developed in PERT. TANES can be used to determine the critical path through a time or cost network and in addition to help establish effective schedules for those paths which are not critical.

The spectacular success of PERT on *Polaris* was instrumental in its being introduced and used in various types of projects other than military ones under its original name or some pseudonym such as

CPM (Critical Path Method), PEP, or PERT/COST.[40] The basic method involves drawing up a diagramatic model of the various tasks to be performed in achieving a given result, assigning estimated times for accomplishing each of the tasks, and determining whether the times required are consistent with the scheduled completion time. If this is not the case, then the critical path of events is determined from the model so that the tasks which are most significant in limiting the realization of the schedule are highlighted.

Presumably, by knowing what constitutes the bottleneck conditions, management is in a position to modify its plans and schedules to alleviate an unsatisfactory situation before it arises. The keys to this method appear to be understanding the necessary steps to be taken to accomplish the job, being able to estimate the range of times that the steps may take, and being willing to modify the steps as the need becomes evident.

Network Analysis Technique

Essentially the network analysis technique consists of the following:

1. Determining end objectives and goals.
2. Establishing a plan based on these objectives.
3. Developing a forecast related to the plan.
4. Structuring in a systematic way activity and decision requirements.
5. Programing the manpower and other resources required.
6. Executing these activities according to the plan.
7. Devising measures for the control of the work.
8. Reporting and evaluating the results.
9. Appraising and reviewing the information for corrective action.
10. Developing preventive action and systems and providing for the improvement of plans on the basis of the feedback of information.

Elements of PERT[33]

The following are the elements on which the PERT concepts are built:

1. *An event.* This is an inexplicitly identifiable point in time at which something has happened or a situation has come into existence. There may be work involved in approaching an event but the event itself takes no time; therefore no work is represented by an event.

2. *An activity.* This is a clearly definable task to which a known quantity of manpower and other resources will be applied. In basic

PERT

PERT, an activity represents effort applied over a period of time and is bounded by two events. These events are referred to as the predecessor and successor events for the associated activity.

3. *Time estimates.* PERT associates an elapsed time with an activity. In order to determine in advance what this time is likely to be, it is necessary to estimate. The estimating procedure is the cornerstone of the PERT technique; someone who is capable of actually performing the activity in question is asked for *three* time estimates.

a. An *optimistic* time: the time which would be required if everything worked out or proceeded ideally. This is an unrealistic estimate to the extent that it can be expected to occur in approximately only one case out of a hundred.

b. A *pessimistic* time: the opposite of the optimistic estimate. Barring totally uncontrollable situations such as fires and floods, it is the time required if everything which could logically go wrong did go wrong. This estimate is also unrealistic, representing the one worst case out of a hundred.

c. A *most likely* time: the time which, in term of the estimator's past experience, this activity is most likely to take in the circumstances expected to exist.

4. *Expected time.* The three time estimates are combined mathematically into two formulas which produce two items of information. The first is the PERT expected time—that is, the time that divides the total range of probability in half. There is a 50-50 chance that the time actually required will be equal to or greater than the expected time. The second is the time spread.

5. *Spread.* Another manipulation provides a measure of the degree of uncertainty associated with the expected time. This measure tells us the width or spread of the center 50 per cent of the total distribution so that we can say that there is a 50 per cent probability that this activity will take the expected time, plus or minus so many weeks.

6. *Network.* In doing a job, the first step is to analyze the component tasks and their interdependencies. The result of this analysis is a network of events and interconnecting activities that defines the series and parallel sequences of activities and events which must occur to achieve the end objective. The second step in doing a job is to secure time estimates and calculate the expected time for each activity. One can then determine the probable length of time required for the various series-sequences of activities which connect the start of the program with the objective event.

7. *Critical path.* One of these sequences will be longer than all

the rest; this longest path is called the critical path because it is the one which determines the length of time required to reach the objective event. It has two principal features. First, if the program is to be shortened, one or more of the activities on this longest path must be shortened or eliminated. The application of additional effort anywhere else in the network will be useless unless the critical path is shortened first. Second, if the time required for the actual performance of an activity on the critical path varies from the calculated expected time, this variation will be reflected in a one-to-one fashion in the anticipated accomplishment of the objective event—no matter how far in the future that event may be.

8. *Slack.* Since the critical path is defined as the longest path in time from the starting event to the objective event, all other events and activities in the network must lie on paths which are shorter. This means that along these paths there is slack, or time to spare. These paths are referred to as slack paths and the areas where surplus resources of men, facilities, or time are to be found.

In order to measure the amount of slack existing at any point in the network, one must determine the earliest expected time and the latest allowable time for each event. The *earliest expected time* for an event is defined as the sum of the expected times for the activities along the longest path leading from the starting event up to the event in question. The *latest allowable time* is determined by adding the expected times for activities on the longest path leading back from the objective event to the event in question and by subtracting this sum from the schedule date for the objective event. The latest allowable time is that time by which an event must occur if slippage of the objective event schedule is to be avoided.

Slack, then, is the difference between the earliest expected and latest allowable times. It represents flexibility, a *range* of time over which the activity can take place without influencing the accomplishment of the objective. Slack areas have not only spare time but also surplus resources of men and facilities.

9. *Probability of success.* A simple arithmetic calculation on the three time estimates enables one to obtain a measure of the uncertanity of the expected time for the activity. Since the expected time for any event is calculated by adding up expected activity—the various times on the longest path leading to that event—one can also statistically combine the uncertainties involved in each activity in such a way as to obtain a measure of the uncertainty in the expected time for the event. Thus, when one calculates the PERT expected time for the end event of a program, he can also obtain a measure

of the *uncertainty* or the range of probable error in the prediction. By another mathematical procedure one may compare the PERT predicted expected time and its uncertainty with the schedule commitment for the objective event and derive the probability of meeting the schedule.

Where to Use PERT

The concept of PERT is broadly applicable in systems engineering as well as in many other fields. Areas where it has been found effective include:

1. Scheduling projects and programs.
2. Evaluating existing schedules as work progresses.
3. Estimating cost for proposed projects.
4. Evaluating cost versus budget as work progresses.
5. Planning resource utilization.
6. Smoothing resource utilization.

Specific projects where the PERT method has been successfully employed are (1) design and manufacture of a new weapons system, (2) implementation of an automated information system, (3) development, design, and tooling for new product models, (4) periodic preventive maintenance or overhaul of a continuous process facility, and (5) design and installation of a new facility.

6.7 Technicalities of Network Analysis[33]

Crucial to the effective use of PERT is the preparation of a valid diagram representing the network of jobs and time which constitutes an effective way of accomplishing the desired project. The systems planner can simplify his task and get better results if he follows certain simple rules. The following series of steps will fit the average case.

1. Define the end objective precisely [Figure 6.7-1(a)]. This is frequently difficult, but close definition is fundamental to a complete plan.

2. Define all significant events that are precedent to the end objective [Figure 6.7-1(b)]. Do not start on any "chain" until this is done. The purpose is, of course, to make sure that no elements are omitted inadvertently.

3. Define all significant events precedent to Event B [Figure

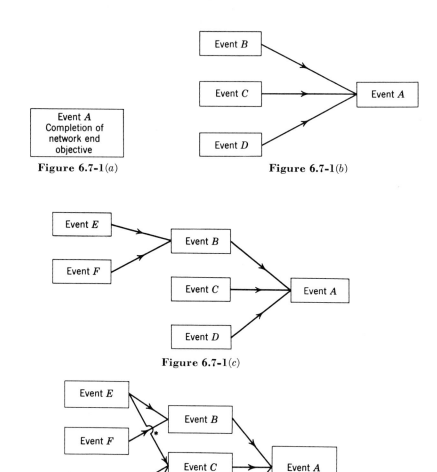

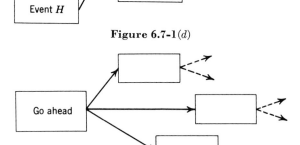

Figure 6.7-1(e)

Successive steps of events in the construction of an activity network diagram.

Technicalities of Network Analysis 249

6.7-1(c)]. This is a continuation of the strategy in the second step. Again, the purpose is to ensure that nothing significant in the whole plan is left out.

4. Define all significant events precedent to Event C. If it is found that some event already shown is precedent to the event being worked on, interconnecting lines must be drawn [see asterisks, Figure 6.7-1(d)].

5. Continue in a similar manner with other events. Work back a level at a time, making sure that all significant precedent events are established. At some point it may be desirable to start afresh at the condition of "Go ahead" and work forward in activities [Figure 6.7-1(e)].

6. Make sure that all events except the beginning and ending ones have at least one connection on each end. Recheck for any important events left out. If there are any, a serious error has been made somewhere. If the omitted event is really important, find out whether it is covered by the scope of the work or is related to it.

In developing the network, it is important to discard irrelevant matters from consideration. There is a tendency to include events that the planner considers necessary, even though they are not essential to the end objective as it has been defined. Although these events may be written down somewhere for reference and follow-up, they should be kept out of the network.

Since it will be convenient most of the time to have the organizations responsible for the various elements of the whole task prepare the portions of the network applicable to their own specialties, a problem may arise in controlling the work sufficiently to ensure a reliable result. It is useful in some cases at least to farm out portions of a top-level network. A simple example is shown in Figure 6.7-2 where the symbols (S) and (C) are used to indicate the start and completion activities respectively. Activities which might be farmed out are A and G to manufacturing; C, D, and E to engineering; B to quality control; and F to shipping. Each would then produce a network for the activities concerned, with the same beginning and end events as on the main network. The whole network, including any organizational points of conflict that may not have shown up on the main network, can then be woven together.

Some managers may want to keep more detailed control over activities in order to avoid undisclosed potentials for exceeding budgeted costs and missing schedules. They may want to know, week by week, what percentage of work is finished as well as what costs are incurred. A few companies have accounting systems that will allow this, but many do not.

A word of caution on the use of percentages is in order. It is essential that system controls be such that misleading figures are avoided. For example, suppose that an activity has an allotted 1000 standard hours, and management wants to measure standard hours completed week by week. At the end of some interval, 500 hours' worth of work, say, may have been reported as completed. Now suppose that the following week an additional 100 hours' worth is completed but that 300 hours' worth of work previously finished is rejected because it was spoiled in a subsequent operation. A poorly designed system might show that 60 per cent (500 + 100 = 600, or 60 per cent of 1000) of activity is finished. Actually, only half that much, at most is now done. It may be even less of a disproportionately large number of parts that will be required to get an acceptable final unit of production.

A better way to measure performance would be to break the activity down into a subnetwork of smaller increments with events whose completion can be individually determined. "Start assembly A" and "Complete assembly A" can be subdivided as shown in Figure 6.7-3.

An interesting problem arises when it is known that the operating plan will be something less than ideal [see Figure 6.7-4(a)]. If the

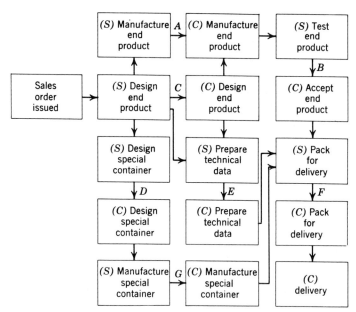

Figure 6.7-2. Representative activity network from sales order issued to delivery.

Technicalities of Network Analysis

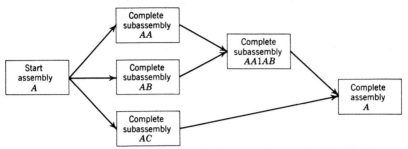

Figure 6.7-3. Subdivision of an activity into a subnetwork of activities can improve clarity.

assembly is an aircraft substructure, for instance, one would know that the given relationship while possibly desirable, is impractical. One would also know that some purchased part can be installed when the assembly is well advanced and that it need not be available before work begins. On an upper-level network the relationships may be represented as in Figure 6.7-4(b). However, for closer control, the situation might be shown in more detail on a lower-level network, as in Figure 6.7-4(c).

The orderly administration of a program using network analysis presupposes careful definition of events. "Complete the design" can mean six different things to six different people. Any communications scheme with this degree of indefiniteness can quickly become intolerable. On a practical basis, the network events have a reasonably definite meaning to those who deal with them every day. On any but a very small scale, however, misunderstandings will occur even within the inner group, to say nothing of those outside this group.

It is advisable, therefore, to define events in writing. When it is feasible, definitions ought to contain the following:

1. Abbreviations used.
2. The long forms of the abbreviated terms.
3. An exact definition of the event.
4. The specific indication or evidence that the event has occurred.
5. The organization responsible for reporting occurrence of the event.

If the same events are included in several networks, the definitions should be standard ones: that is, "complete the design" should have the same meaning in all networks.

The occurrence of events should be signaled by substantial evidence other than the completion of activities, and this evidence should be included in the definitions. For example, the event "(S) Pack for delivery" in Figure 6.7-2 depends on three others: "(C) Manufacture special container," "(C) Prepare technical data," and "(C) Accept end product." Affirmation that these three are completed is not evi-

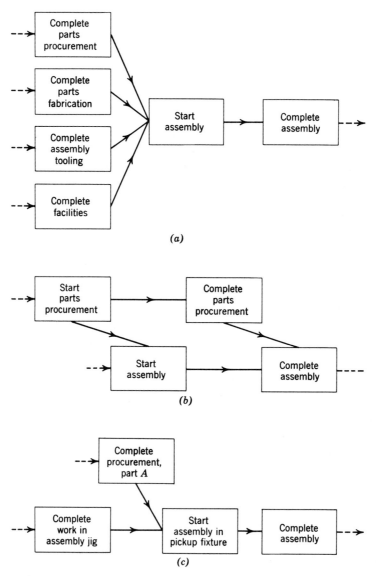

Figure 6.7-4. Clarification of network by more detailed event representation.

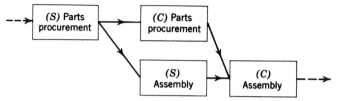

Figure 6.7-5. Network showing installation of the last part in the assembly.

dence of the occurrence of the event "(S) Pack for delivery." This point can become very important. For general applications, network analysis is always event-oriented.

Activities on a network will have to be named or described—that is, one must pay attention to not only the nature of events but also the nature of connecting activities. Defining these activities can result in a useful inquiry into the elements of the plan. Consider the example in Figure 6.7-5. What is the nature of this activity? It can be correctly deduced as the installation in the assembly of the last part to be received, and the time ascribed would then be the time for this assembly.

In general applications, however, it is desirable to ignore the nature of activities in the initial construction of the network. When one delves into activities, he will probably find it necessary to insert a few more event points to take account of, say, changes of responsibility. If "(C) Assembly" involves final inspection, it might be improper to include inspection in this activity; then an additional event point will be needed. All this offers no particular difficulty and is definitely helpful in thinking out the plan.

Whether single "most likely" times or sets of "optimistic, most likely, and pessimistic" times should be assigned to activities will be determined by circumstances. In some cases a single time for activity will be most appropriate: the likelihood of deviation is not substantial. There may be no deviation because management will take whatever steps are necessary to maintain schedule. (This is not a wholly valid argument because management might seek information on the *extent* of the risk to schedule.) In any event, if a computer is not available, it will simplify things if single times are used.

Where a program is on less solid ground and where probabilities are less sharply limited, sets of times may be preferable. The estimated time is usually calculated from the set of times by using a formula that is based on a premise reasonably suitable in most but not necessarily all instances; it may be found advisable to modify

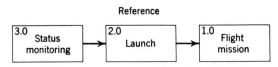

Figure 6.7-6. Reference level functional flow diagram.

this formula in particular cases to obtain a better approximation of reality. This will depend upon the circumstances, and no fixed rule is needed.

In preparing the network or functional flow diagram it is desirable to establish a proper order of levels for the events or activities. The term level describes the degree of detail provided in the diagram in accordance with the coarse, medium, and fine references to modeling given in *Systems Engineering Tools*.[49] By having activities and/or events of comparable levels, one can establish a consistent set of factors of importance for comparison and evaluation. The following example from *AFSC Manuals*[43] 375/310 serves to illustrate the concept of levels.

Examples of an Activity at Different Levels

Space system X—obtains from top-level flow diagram description 1.0 flight mission;

　A. Deliver payload into window at speed vector.
　B. Use 1½-stage liquid-fueled vehicle.
　C. Stage 1 and stage 2 to provide thrust simultaneously until stage 1 shutdown.
　D. Astronaut to have manual staging override capability.

Consider Figure 6.7-6 to show a reference level functional flow diagram for a portion of the time sequence of events in a space system.

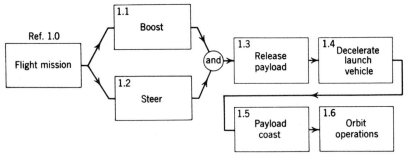

Figure 6.7-7. First-level functional flow block diagram of flight mission 1.0.

Task Network Scheduling

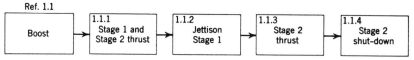

Figure 6.7-8. Second-level functional flow block diagram of boost 1.1.

Flight mission 1.0 is shown to be one major activity and includes the four major attributes listed.

A more detailed description of the flight mission is shown in the first-level flow diagram of Figure 6.7-7, where reference level activity 1.0 is shown to be equal to a combination of activities 1.1–1.6.

A still more detailed flow diagram at a second level is shown in Figure 6.7-8, where boost activity 1.1 is shown to be composed of the four activities 1.1.1 through 1.1.4. In similar fashion each of the other items, 1.2 through 1.6, of the first-level activity diagram, Figure 6.7-7, should be represented by an appropriate second-level flow diagram similar to Figure 6.7-8. Obviously, still higher-level diagrams may be called for, showing in even further detail the individual activities and time elements involved. It is highly desirable that a consistent degree of detail, i.e., comparable bases for comparison, be shown in the flow diagrams at each level. In general, it does not pay to go to too high a level of diagram before the activities and events on the lower levels are found to be compatible with the reference diagrams for the other portions of systems which affect a particular activity in a major fashion. Occasionally, however, consideration of still higher levels of activities is required to bring out important details which may be crucial to successful functioning at lower levels.

6.8 Task Network Scheduling

A number of the steps in the process of Task Network Scheduling (TANES) will be brought out by consideration of a particular problem. Figure 6.8-1 illustrates a small network of the complex kind requiring TANES. In reality, the complexity of networks which may be handled by TANES is also amenable to PERT. The presentation of TANES which follows serves to show in a more quantitative fashion some of the detailed problems arising in the use of networks for scheduling. In Figure 6.8-1 there are far fewer tasks (only eighteen) than in many real projects, which may have several hundred or perhaps

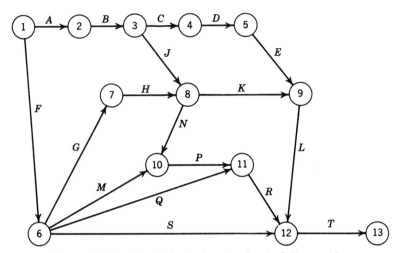

Figure 6.8-1. Illustrative task network scheduling problem.

a couple of thousand tasks if the higher levels of details are included. The relationships among the tasks in the figure, however, have the characteristic complexity of a network needing TANES. The following material draws heavily from unpublished notes of Dr. Wallace Barnes.

In Figure 6.8-1 each lettered arrow represents a *task* or activity, and each numbered circle represents an *event*. An event is a point in time when one or more tasks may begin and/or when one or more tasks must have been completed. Thus, event 3 is the completion time of task B, and also the earliest time at which tasks C and J can begin. It will be noted that there is a unique first event, labeled 1, which precedes *all* tasks. There is also a unique last event, labeled 13, which *follows* all tasks. This is a characteristic which must exist in order to use TANES. In some cases, dummy first and last events must be introduced.

The task network establishes all the precedence relationships which the *technology* of the project requires. Thus, for example, task H cannot begin until task G has been completed. But task H is independent of task M and of task B. Task K cannot begin until *both* task H *and* task J are complete. Also, task K must be preceded by tasks B and G, through implied precedence relationships. A listing of all tasks, together with all *immediate* predecessors for each, would be equivalent to the network diagram. However, it is usually easier for operating department personnel to examine a diagram, such as

Task Network Scheduling

Figure 6.8-1, in checking for omissions and inconsistencies. It should be noted that arrows should *not* be inserted to indicate precedence relationships dictated by *resource availability,* and not technology.

The feature of the network in Figure 6.8-1 most characteristic of the kind of network where TANES is needed is the interconnection of horizontal task levels. Tasks such as *J* and *N* so complicate the scheduling of the project as to make much more difficult the use of methods other than TANES, especially if 500 or more tasks are related in this fashion.

The problem solved by TANES is to set scheduled completion dates for all tasks, given the precedence relationships implied by the network, the due date for the end event, the starting date for the first event, and the time required to perform each task. For each task, the time required can be either a single variable or a number calculated from a probability distribution of possible times for each task. Generally, the probability distribution is used by stating the "most likely," "optimistic," and "pessimistic" times for performing each task, assuming a fixed intensity of resource utilization, e.g., a fixed number of men working on the task. As explained previously, the "most likely" time to do the task is the time which has a 50-50 chance of being exceeded. The optimistic time is the time which has a chance of only 1 in 100 of being improved upon. The pessimistic time is the time which has only 1 chance in 100 of *not* being beaten by actual task time. These three time estimates are combined in TANES into a single time estimate, called the *expected time,* and a *standard deviation* in expected time. "PERT Summary Report, phase 1"[14] explains how the expected time and standard deviation in expected time are computed.

One other note should be made about the typical TANES network. Often some of the tasks are dummy activities. They require zero time and are entered in the network solely to preserve the proper precedence relationships. On Figure 6.8-1, for example, it might be that task *K* could begin as soon as tasks *H* and *B* have been completed. In this case, task *J* is a dummy. It takes zero time to perform task *J.* The network in Figure 6.8-1 is still correct. The only reason for introducing *J* is to ensure that *B* is complete before *K* begins. (It should be noted that it would not be the same network if *B* ended at event 8. Then *C* would begin after event 8, and it would be implied that *C* also must follow *H.* This is not the case in Figure 6.8-1.) Sometimes dummy tasks are indicated by broken-line arrows. The use of dummy tasks and dummy events is an important part of the technique of developing TANES networks. Re-examination of the network with operating personnel familiar with technological precedence requirements is usually a vital first step in using TANES. The first network drawn will almost always have to be modified as a consequence of this re-examination. Thus, a soft pencil and a reproducible mat are recommended for drawing the network.

Network Analysis

The first step in any of the task-scheduling techniques is to analyze the precedence relationship network to find the amount of *float* (or "slack" in PERT terminology) available on each *chain* of tasks in the network. Float is the difference between the time available to do a chain of tasks and the sum of the expected times of the tasks on the chain as computed from the input task time estimates. Here a *chain* of tasks is a set of tasks which follows one another in a network, such as tasks *A, B, C, D,* and *E* on Figure 6.8-1. A chain may or may not go all the way from the first event to the last event on the network. However, the term *path* is used to describe a chain, such as *A, B, C, D, E, L, T* on Figure 6.8-1, which *does* go all the way from the first event to the last event.

The important fact about chains, or paths, of tasks is that in the TANES network analysis the float which is found first is the one which must be *shared* by all the tasks on the chain. Thus, for example, if chain *A, B, J* must be completed by week 10, and cannot be started until week 3, then the available time for the chain is 7 weeks. Now suppose that the sum of the expected times of tasks *A, B,* and *J,* as computed from input to the TANES program, is 5 weeks. Then chain *A, B, J* has 7 minus 5, or 2, weeks of float. The 2 weeks may be *allocated* among the three tasks in any fashion desired, but the sum must be 2 weeks. In some TANES versions, notably in PERT, the float is *never* allocated. PERT does *not* develop a schedule for the network. However, PERT does find the float for the various paths in the network. It uses this float, then, to evaluate a schedule which is provided as input; that is, PERT finds the *probability* that a given schedule will be met, considering up-dated information on precedence relationships and task time estimates. CPM (Critical Path Method, which is somewhat related to PERT), on the other hand, *does* allocate the float along each chain and produces a schedule for the project. But CPM computes no probabilities. This difference is the principal one which distinguishes PERT from CPM.

All TANES programs, then, begin by determining the *path* float for each *task* in the network. To do this, one must first sort out

Task Network Scheduling 259

all the pertinent paths in the network and then find the float for each path. The path with the least float is called the *critical* path of the network. All the other paths are then ranked by path float, with least path float first, and *chains* are set up by deleting from each path all the tasks already assigned to a path of smaller path float.

This scheme of network analysis is simple, at least in principle. But in a complicated network of 500 or more events, the identification of all the pertinent paths could be difficult. Fortunately, a simple device not only identifies the appropriate paths but also generates information useful to a limited extent for its own sake. This device is the computation of an *earliest completion date* and a *latest completion date* for each task in the network. The difference between these two dates turns out to be the path float for the path of least float involving the given task.

Example: The principle of network analysis can be seen most easily by working out an example. The project here is to install an EDP (Electronic Data Processing) system in the Hypo Department (Hypo for Hypothetical). The tasks required to install the EDP system have been developed by the operating personnel of the Hypo Department. While entirely hypothetical, this example was distilled from actual experience in using PERT. It is simplified for ease in understanding.

In Table 6.1-8 the column labeled "Expected Time Required" is the anticipated applied time duration, in weeks, to do each task. Following the best TANES practice, these times were obtained by asking operating personnel for most likely, optimistic, and pessimistic times, and then averaging them. For example, the times to do a preliminary systems design were originally estimated as follows:

Most likely	5 weeks
Optimistic	4 weeks
Pessimistic	12 weeks

Using the recommended weights of $\frac{2}{3}$, $\frac{1}{6}$, and $\frac{1}{6}$ on these three estimates yields the average estimate of 6 weeks shown in Table 6.8-1. The last column in this table shows the resource requirements, in number of men, of type A, B, C, or D. Type A are systems designers; type B are men with finance background; type C are coders; type D are computer programmers. Thus, preliminary design requires three systems designers, etc. It will be noted that some tasks, e.g., "Obtain delivery of computer," require time but no Hypo Depart-

Table 6.8-1. *Tasks Required to Install an EDP System in the Hypo Department*

Description of Task	Expected Time Required (wk)	Resource Requirements	
		Quantity	Type
1. Do a preliminary systems design	6	3	A
2. Prepare specifications for computer	3	1	B
3. Select computer and place order	4	2	B
4. Obtain delivery of computer	26	...	...
5. Design computer site	6	2	B
6. Write contract for site development	4	1	B
7. Develop site	13	1	B
8. Check out computer operation	4	2, 1	B, D
9. Write appropriation request	6	2	A
10. Get approval of appropriation request	13	...	...
11. Do detailed systems design	10	3	A
12. Write and code computer program	4	1	D
13. Convert a sample of input	3	1	C
14. Complete data conversion	12	2	C
15. Test and modify program	2	1, 1	A, D
16. Check out system (computer, data, program)	6	$\frac{1}{2}$	D

Resource key: A = Systems designers; B = Finance men; C = Coders; D = Programmers.

ment resources. One task, the last one, requires a fraction of a man.

The next step is to draw the precedence relationships among these tasks. When this is done, and all the important dummy precedence relationships have been determined, it just *happens* that the network looks like Figure 6.8-1. Looking back, one can identify each task as shown in Table 6.8-2.

The remaining tasks shown in Figure 6.8-1 are the dummies inserted to preserve proper precedence relationships. They are as follows:

Reason for Inserting Dummy	Task Letter	Time Required (always 0 weeks)
Approval of appropriation request (B) must precede Obtain delivery of computer (K)	J	0
Select computer (H) must precede Write and code computer program (P)	N	0

Task Network Scheduling 261

Table 6.8-2. *Task Identification from Figure 6.8-1 and Table 6.8-1*

Task Identification	Task Letter	Time (wk)
Do a preliminary systems design	F	6
Prepare specifications for computer	G	3
Select computer and place order	H	4
Obtain delivery of computer	K	26
Design computer site	C	6
Write contract for site development	D	4
Develop site	E	13
Check out computer operation	L	4
Write appropriation request	A	6
Get approval of appropriation request	B	13
Do detailed systems design	M	10
Write and code computer program	P	4
Convert a sample of input data	Q	3
Complete data conversion	S	12
Test and modify program	R	2
Check out system	T	6

The first step in applying TANES is *network analysis*. The critical path, and all the other paths, on Figure 6.8-1 (with the applied times as stated in the tables above) must be identified. To do this, the earliest completion date and latest completion date for each task are determined. Before doing this, however, it is helpful to consider renumbering the events which begin and end the tasks in the network. The renumbering is not required, but if a certain rule is followed the latest and earliest date schedules can be computed in a more straightforward manner. This rule is that no event shall be numbered until all its *predecessor* events (the beginnings of the tasks leading into the given event) have been numbered. This still leaves many possible event numberings, and all are equally acceptable. But it rules out, for example, on Figure 6.8-1, assigning event 6 to the end of task *E*, because a task arrow (letter *K* ends at this time and the *beginning* of arrow K has not been labeled with a number less than 6. Similarly, the event at the end of tasks *L*, *R*, and *S* must be left unnumbered until the beginnings of all three of these tasks have been numbered. Applying this rule to Figure 6.8-1, one finds that the given numbering of events is acceptable and no renumbering is necessary.

The numbering of events in this fashion allows a very easy method for listing all the tasks in the network for earliest date and latest date scheduling. Each task has a successor *event* and a predecessor *event*. The method of listing the tasks is to write the *successor* events in numerical order, and where several tasks have the same successor to list all the predecessor events in sequence. In other words, the *major* sort is on successor event and the *minor* sort is on predecessor event. Applied to Figure 6.8-1, this method produces the information in Table 6.8-3.

Here it will be observed that the first item written down is the *successor* event, in column 2, and then the *predecessor* event is entered in column 1. The PERT version of TANES reverses these first two columns, and labels task *A*, for example, as tasks (2, 1) instead of (1, 2). This seems a bit awkward for manual manipulation.

In this table, the task applied times were copied in, and then the

Table 6.8-3. *Task Event Numbering Showing Predecessor and Successor Events and Other Significant Dates for Table 8.6-2*

Task No.		Task Applied Time (wks)	Earliest Completion Date	Latest Completion Date	Latest Start Date	Path Float (wks)
Pred. Event	Succ. Event					
1	2	6	7	12	6√	5
2	3	13	20	25	12√	5
3	4	6	26	34	28√	8
4	5	4	30	38	34√	8
1	6	6	7	18 39 43	37√	11
6	7	3	10m	21	18√	11
3	8	0	20m	25	25√	5
7	8	4	14	25	21√	11
5	9	13	43	51	38√	8
8	9	26	46m	51	25√	5
6	10	10	17	49	39√	32
8	10	0	20m	49	49√	29
6	11	3	10	53	50√	43
10	11	4	24m	53	49√	29
6	12	12	19	55	43√	36
9	12	4	50m	55	51√	5
11	12	2	26	55	53√	29
12	13	6	56	61	55√	5

Project due date: Week 61.
Project start date: Week 1.

Task Network Scheduling

earliest due dates were obtained by going straight *down* the table, without looping or skipping any rows, while latest due dates were determined by going straight *up* the table. A project due date of week 61 has been assumed. Wherever a successor event appears more than once, a small "m" has been inserted to show the *maximum* earliest due date for use with dates scheduled *after* the maximum choice. Thus, for example, the earliest due date for event 10 is the maximum of 20 and 17, or 20. Thus, task (10, 11) is scheduled for 4 + 20, or week 24.

The easiest way to obtain latest due dates is to start with the last task (12, 13), assign to it the project due date as latest due date, and then subtract task applied time to get the latest *start date* for (12, 13). This start date is *tentatively* entered as the latest due date for all taks which have task 12 as *successor* (second column). If at any time a *smaller* entry for one of these tentative *due* dates is obtained from some other successor, this smaller entry replaces the tentative latest due date. For example, after the latest start date for task (12, 13) is set at week 55, a 55 is set into the latest due date for all tasks ending at 12, in this case for tasks (6, 12), (9, 12), and (11, 12). By subtracting task applied time, this gives a latest *start* date for (6, 12), for example, of week 43. This is set as the tentative latest due date for all tasks ending with 6 [in this case task (1, 6) is the only one]. Later on in the backwards scheduling process, the latest start date for (6, 11) is set at 50. This 50 is then also tried for the due date of (1, 6). If *smaller* than the inserted 43, the 50 would replace the 43. When the 50 is not smaller, as is the case here, the 43 stays in. Later in the scheduling process, however, a latest start date of week 39 is set for task (6, 10). This 39 must then be tried as a tentative latest due date for task (1, 6). Since 39 *is* smaller than 43, the 43 is replaced by 39. Later still, the 39 is replaced by an 18. For ease in following the construction of latest due dates, the crossed-out dates in Table 6.8-3 have been left showing. As each latest start date is obtained, from the bottom upwards in the table, it should be tried as a due date for predecessor tasks. (A check mark in this column helps keep track of start dates transferred as tentative due dates.)

After the path floats have been obtained by finding the difference between the latest and earliest due dates in the last column of Table 6.8-3, path float sorting gives the data shown in Table 6.8-4.

In this table, the *chains* are shown by the *underscored tasks* placed next to one another, while the additional tasks from more critical paths needed to turn each chain into a path (i.e., make it go from

Table 6.8-4. *Path Float Sorting for Data in Table 6.8-3*

Path No.	Path Float (wks)	Path Tasks	Tasks Req'd to Complete Chain
1 (Crit. path)	5	(1, 2) (2, 3) (3, 8) (8, 9) (9, 12) (12, 13)	None
2	8	(1, 2) (2, 3) <u>(3, 4) (4, 5) (5, 9)</u> (9, 12) (12, 13)	#1 (1st, 2nd)* #1 (5th, 6th)
3	11	(1, 6) (6, 7) <u>(7, 8)</u> (8, 9) (9, 12) (12, 13)	#1 (4th, 5th, 6th)
4	29	(1, 2) (2, 3) (3, 8) <u>(8, 10) (10, 11)</u> (11, 12) (12, 13)	#1 (1st, 2nd, 3rd)
5	32	(1, 6) <u>(6, 10)</u> (10, 11) (11, 12) (12, 13)	#3 (1st) #4 (5th, 6th) #1 (6th)
6	36	(1, 6) <u>(6, 12)</u> (12, 13)	#3 (1st) #1 (6th)
7	43	(1, 6) <u>(6, 11)</u> (11, 12) (12, 13)	#3 (1st) #4 (6th) #1 (6th)

* That is, the first and second tasks on path #1, or tasks (1, 2) and (2, 3).

event 1 to event 13) are shown without underscoring. Thus, path #2 consists of a chain of three tasks, (3, 4), (4, 5), and (5, 9), plus two tasks from the beginning of path #1 and two tasks from the end of path #1. In selecting the additional tasks to turn a chain into a path, it is essential to choose them from the path of *lowest* path float. Thus, for example, in completing path #4, a set of tasks going from event 1 to event 8 is part of the requirements. There are two sets available. They are part of path #1, which goes from event 1 to event 8, and part of path #3, which also goes from event 1 to event 8. The tasks selected to complete path #4 *must* be selected from path #1 because this path has the smaller path float, and hence the longer task applied time in going from event 1 to event 8. The last column of Table 6.8-4 tells the path number, and the task number on this path, required to complete each chain and turn it into a path. Thus, path #2 consists of chain 2 plus tasks 1 and 2 of path #1, plus tasks 5 and 6 of path #1. As a check, it should be noted that the number of tasks assigned to all chains must equal the total

Task Network Scheduling

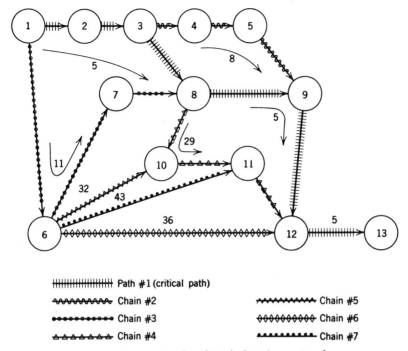

Figure 6.8-2. Analyzed typical project network.

number of tasks in the network. In this case, this is 18. There are 18 underscored tasks in the "Path Tasks" column.

This completes the network analysis in this example. The analyzed network is shown in Figure 6.8-2. Another method of drawing the analyzed network, which displays more effectively the pertinent chain for each task, is shown in Figure 6.8-3.

Thus the critical path includes the tasks of writing the appropriation request, getting it approved, obtaining delivery of the computer, checking out the computer operation, and checking out the system. Two of these tasks are out of the control of the Hypo Department. However, there are two other paths within a few weeks of being critical. One path includes all the tasks associated with computer site development; the other path includes preliminary system design and ordering the computer. The Hypo Department might very well concentrate its attention on these tasks. On the other hand, it could easily afford to delay the detailed systems design, programing, and data conversion. This illustrates the kind of information which a TANES analysis provides to systems management.

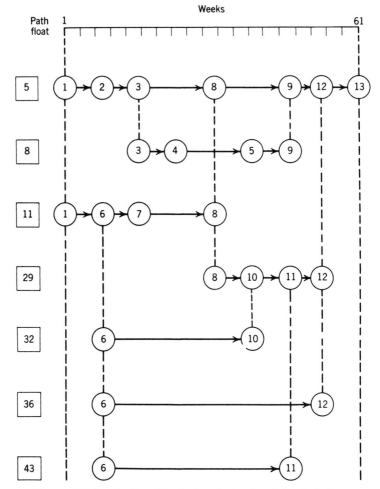

Figure 6.8-3. Analyzed typical project network in chain form.

Project Schedule Evaluation

The objectives of using a TANES program are often broader than just project scheduling. Another important application of TANES is the evaluation of a schedule which already exists, often in the form of milestones on the network. These milestones are key events for which completion dates have already been specified. Many times the customer, especially in the defense business, specifies the date for each milestone. The question to be answered is then, "What is the probability of reaching each milestone by the given date?" This is the sort of question which PERT (Program *Evaluation* and

Review Technique) was really designed to answer. PERT was not designed to schedule a project.

How are the probabilities to be computed? The input information is the same as in the basic TANES concept, i.e., a network and task times, *plus* the given milestone dates. The network is analyzed into paths, and the critical path from the beginning event to each milestone is determined. The earliest completion date for each milestone is then computed and compared with the given date for that milestone. The difference is expressed in units of the standard deviation in task time along the critical path, and a table of the normal probability distribution is used to compute the probability of meeting the milestone date.

The problem of estimating the standard deviation in task time along the critical path is recognized as difficult. PERT does this by assuming statistical independence among task time estimates along a path and by ignoring variability in estimates on subcritical paths. An alternative approach might be a Monte Carlo analysis. Even the strongest advocates of PERT have expressed doubt about the statistical validity of the assumptions made. But there is no doubt about the need to be able to evaluate the probability of meeting a milestone date. Also, the probabilities found by PERT have, in practice, been good enough to alert the Navy's Fleet Ballistic Missile (FBM) management as to where its contractors were likely to stall the FBM program.

Project Cost Estimation and Cost Evaluation

Another application to which task network scheduling has been put is to generate estimates of the cost for a project, and to evaluate the probability of overrunning the budgeted cost for a project in progress, sometimes called PERT/COST.[40] This application requires the same input as the basic TANES concept, plus an estimate of the cost for each task in the network. It is even possible to specify alternative costs which might be incurred if the time to perform the tasks were altered. In this way, TANES can be used to convert cost-time trade-offs for each task into cost-time trade-offs on the complete project. In the TANES algorithm used in this refinement, provision is made to shorten only those tasks which lie on the critical path. Even on the critical path, the tasks with smallest cost gradient are shortened first.

Dynamic Modeling and Simulation for Time and Cost

The preceding material, which describes the PERT and task network scheduling approaches to understanding and predicting the likely occurrence of activities with time in the future, has employed a range of time estimates of events derived from previously acquired statistical data. The same structural flow diagram approach, or one somewhat modified from it, can be employed perhaps to even greater advantage at times by using the dynamic modeling and simulation approach of industrial dynamics pioneered by J. W. Forrester[26, 46] and others at M.I.T. and the application of feedback control techniques to organizational systems set forth by R. N. Wilcox.[35] In these approaches the conventional input-output techniques of block diagrams and transfer functions are combined with the digital logic and probability of statistics to provide a means of relating them to the likelihood of certain different activities occurring in concert. Thus we are able to consider what will happen after event A if the result of A has been successful or unsuccessful; in one case we will go to event B, in the other to event C. Monte Carlo[49] or more conventional simulation methods can be used to evaluate models of these kinds. Such a detailed and flexible portrayal of activities and events represents a logical and useful extension of PERT and TANES concepts and is finding ready acceptance in time and cost control and management of organizational systems,[46] including those required to build systems.

6.9 Conclusions

Time is a crucial ingredient in the lives of men and in the conduct of activities. It is essential in systems engineering that careful attention be given to the time that is required to make the system as well as the time that the system itself takes to operate. In recent years management control techniques such as PERT and TANES have been developed which make it possible to handle in a systematic and organized fashion the prediction and monitoring of the time to accomplish system-type projects as well as other significant events. These methods are very beneficial in helping the systems engineer to keep track of time and to update his plans for the handling of his project as real events occur and replace his original estimates of the time required for various activities.

7

Reliability

7.0 Introduction

The term reliability has become increasingly important in recent years.[32, 37, 38] The growth of increasingly complex military systems and their vital role in the defense of our country have focused attention on the reliability of such systems. The automatic control of large and expensive industrial processes has served to emphasize the importance of reliability as a performance factor comparable with such other requirements as efficiency, speed, and accuracy. In the domestic area, reliability is a factor of definite marketing value in ensuring the continued sales of appliances. The fact that a system or device is not operating is an observable condition of which even the least sophisticated of individuals can be made aware.

In the material which follows, a brief description is given of what is meant by the term reliability and what are some of the many things being done to achieve it. Not only are engineering, manufacturing, quality control, and maintenance concerned with reliability, but also management itself at all levels from the top down is devoting increasing attention to this requirement.

Although reliability arithmetic sometimes is made to appear as if it constituted the entire reliability problem, this subject is presented here only in sufficient detail to illustrate the nature of some of the problems.

The principal emphasis is placed on the many facets of the problem of designing for reliability. Stress requirements, establishing the preliminary design, identifying the failure modes, determining the effects of tolerances and parameter changes, and the use of tests and design reviews for bringing about better reliability are described. Later quality control methods for improving reliability are discussed. Maintenance is likewise presented as a factor for helping to make a system continue to perform reliably over longer periods of time.

Finally, the role of management in providing an environment of personal leadership which fosters rather than hinders reliability is reviewed briefly. This chapter on reliability is a long one, but the importance of the topic warrants this emphasis.

7.1 Reliability and What Is Being Done About It

Reliability has been defined as *"the probability of a device or system performing adequately for the period of time intended, under the operating conditions encountered."*[5] This definition includes such diverse cases as that of a motor which must start and stop, and run perhaps for years without failure, and that of a ballistic missile system which must be in standby condition for long periods of time and then must respond just once—but once without fail. In all cases it can be determined that success in meeting the reliability requirements will be critically dependent on the extent to which reliability is designed and built into the product.

Reliability is an analytical problem involving both statistical and engineering aspects. It must be given critical attention throughout the life of a job—in development, design, production, outgoing quality inspection, shipping, installation, operation and maintenance. It requires the integrated application of many disciplines—statistics, materials engineering, circuit analysis—all the engineering disciplines, in fact, plus sound organization for seeking and correcting the causes of failure. Component reliability is a condition necessary but not sufficient for systems reliability.

What Is Being Done about Reliability?

The problem of promoting reliability has always been present but has increased in recent years with growing system complexity. Since 1950 and the Korean War, the search for reliability has been given the status of a special endeavor, and considerable progress has been made in various facets of the problem. This came about largely because of the obvious need for reliability in military hardware and because of the insistence of the industrial user on automatic control for reliable systems. It is now recognized that the techniques and procedures worked out to ensure the performance of military equipment can and must be applied elsewhere. On the other hand, it is being more fully appreciated by the designers of military equipment that some of the design techniques employed in industrial equipments can be used to advantage in improving military design methods.

Basically, the approach to reliability has been to admit that it presents a problem which has a number of solutions and to organize to find these solutions. Some of the specific things that have been done to improve reliability include the following:

New engineering and other sections have been brought into being and charged with reliability, responsibility, and authority.

The nature and causes of failure have been sought, and facts have been obtained.

Increased emphasis has been placed on component reliability, and marked improvements have been made in the design and manufacture of components with reliability as a major objective.

Statistical methods have been explored extensively to gather data and to analyze how they may be applied to the synthesis as well as the analysis of reliable equipment.

Nationwide technical conferences devoted exclusively to reliability problems have been held annually for a number of years.

Reliability design objectives have been set forth in many contracts, and many engineering designs have been analyzed for their inherent capability in this respect.

Manufacturing methods, including quality control and test of incoming material, have been more critical and searching in an effort to weed out materials, components, devices, and equipment which fail to meet the design specifications.

Systems approaches to reliability have pointed out the need and indicated ways for using redundancy to improve reliability, have emphasized the necessity for adequate maintenance and servicing, have helped determine the extent of the spare parts requirements, and in general have tended to stress the overall requirements for reliability in operation.

The mere recognition that securing equipment reliability can be a problem does not in itself provide a solution. *Effort* in the form of well-directed reliability-oriented programs is the key to measurable achievements in this area.

Reliability from a Period-of-Time-Intended Viewpoint

The concept of reliability (the probability of a device performing adequately for the period of time intended) provides an insight into the reliability problem and a clue to how it may be handled successfully. The strength of any practical object will be greater than the load applied (or stress) for only a finite time because of the deterioration which occurs to the object over a period of time. For the object not to fail, it is necessary for its strength to exceed its stress.

Figure 7.1-1 illustrates in simplified fashion the nature of the prob-

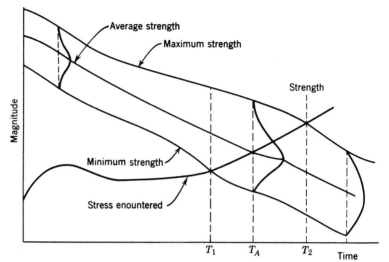

Figure 7.1-1. Strength and stress encountered as a function of time.

lem. Shown on the upper curves as a function of time for a given object is the range of its strength. The term average strength is used to indicate that not every one of these objects has the same strength at the same time. This average strength is a gradually decreasing function with time as a result of such factors as insulation deterioration, corrosion, mechanical fatigue, or other aging effects. The curve starting at the lower left shows the stress encountered by the object, also as a function of time. The varying stress is a result of different environmental operating conditions encountered. In many cases it is a problem of no small magnitude to establish just what the environmental conditions are. What is shown as a single stress-encountered curve may itself have a range of values. On the average, for time less than T_A the strength exceeds the stress, and the object will perform satisfactorily. Failure should occur at time T_A under normal conditions.

As is generally recognized, not all objects subjected to the same stress will fail at the same time. A number of probability distributions about the average strength curve are also shown. These distributions indicate that, at any particular time, there is a range of values of strength about the average with some frequency distribution as shown. Connecting the minimum strength points and determining the intersection of the stress line with this locus of points of minimum strength permit determination of time T_1, at which failure is first apt to occur. In similar fashion, from the points of maximum strength, time T_2 at which failure is last likely to occur can be determined. Thus, using the definition of reliability as a probability, one obtains a curve as shown in Figure 7.1-2 for reliability as a function of time. Although some few failures may occur at times shorter than T_1, it is apparent that good design practice will require that the "period of time" for which reliable use may be expected for the object considered must be less than T_1.

Many factors may affect the reliability picture presented so as to alter the characteristics shown in the graph denoting strength versus time: the stress encountered may differ from the values shown, the average strength may be altered as a result of the stresses that are applied, and the probability distributions may be changed. However, the same concept of reliability being a decreasing function of time which is dependent on strength and stress is valid.

In another way of looking at reliability the failure rate as a function of time is studied. Figure 7.1-3 shows that for the initial period of time after the equipment is placed in operation there may be some early "infant-mortality" failures up to time T_c. Factory tests and run-in operation periods may do much to prevent these failures from occurring in actual normal operating use.

From T_c to T_1 the failures are relatively few and are frequently considered to occur at a constant failure rate, hence the name for this interval. The greatly increased failure rate starting at T_1 has been described and is frequently associated with the term wear-out. These three intervals have been named in a qualitative fashion; no significant quantitative values are associated with the terms, and these will vary from system to system. This curve is sometimes referred to as the "bath-tub" characteristic.

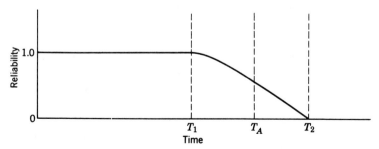

Figure 7.1-2. Reliability versus time.

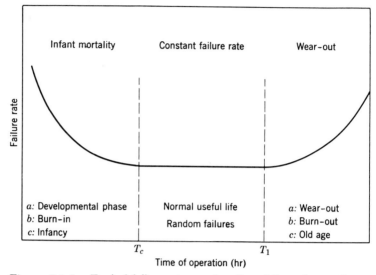

Figure 7.1-3. Typical failure rate as a function of time of operation.

Reliability under the Operating Conditions Encountered

The reliability definition states that the device performs adequately "under the operating conditions encountered." These operating conditions are, in effect, the environment under which the equipment performs in its assigned role. They are the stresses encountered in the first graph. These stresses must be understood completely by the designer so that he may incorporate in his design the inherent strength to ensure the ability of the device to meet them. Sometimes the stresses are spelled out satisfactorily in the specifications; sometimes they must be determined.

In addition to a knowledge of the actual operating conditions and environment, the preoperational environment from fabrication, through assembly, quality control and test, shipment, and installation, must be understood and incorporated into the stress-encountered picture. Since these stresses can in fact determine or alter the average strength as well as its probability distribution, as described in Figure 7.1-1, the reliability plot shown in Figure 7.1-2, which is determined from these data, can be markedly altered.

Skill, knowledge, and judgment are required to establish a thorough understanding of the operating and preoperating conditions to be encountered by a device. Figure 7.1-4 shows a schematic block diagram

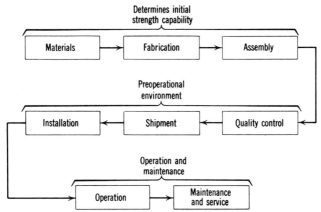

Figure 7.1-4. Functional factors influencing environmental conditions and device capability.

of the major functional factors influencing the environmental conditions which a device of a given design experiences. Thus, not only must the equipment be designed so as to operate as required; it must also be able to withstand all the conditions which it encounters during the preoperational environment as well as those that are a part of its being made.

Referring to Figure 7.1-4, the areas described as materials, fabrication, and assembly establish the initial strength and capability of the device. Quality control and test, shipment, and installation form the preoperating environment, which may alter the strength capability with which the operational life is entered. Operation and maintenance and service represent the actual operating environment as influenced by the effect of such factors as preventive maintenance and preoperational confidence checking. All these elements in varying degrees and at different times are influential in determining whether the device performs adequately "under the operating conditions encountered."

Life Model

As a means of better appreciating the effect of various phases of the reliability problem, it is convenient to have in mind a model or visualization of the overall life of a device. Figure 7.1-5 shows such a pictorial representation from the research and development initiation of the basic concept to the field operation, maintenance,

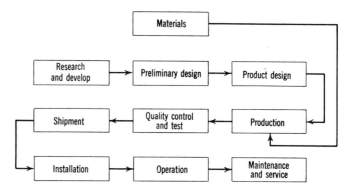

Figure 7.1-5. Device life model.

and service years later. The diagram indicates that the research and development and the preliminary and product designs must have as part of their constraints the way in which the device will be built and used. Furthermore, the steps involved in specifying the incoming materials and fabrication methods indicated in Figure 7.1-4 must be included as part of the overall reliability consideration.

Although the device life model will be discussed later in more detail, a number of very fundamental generalizations are fairly apparent from this figure.

1. Many people and functions contribute to the reliability of a device. No one person or group can alone ensure reliability. Conversely, everyone must be reliability conscious and must work toward improvement in this respect.

2. Because of the high degree of interdependence of various groups in achieving overall reliability success, a good system of communications for feeding information forward and feeding it back is necessary for high reliability.

3. The environmental conditions of operation and maintenance must be established at the outset to the greatest extent possible so that the initial preliminary design and product design are compatible with the actual environment the equipment is to encounter.

4. Since rejection or failure of a device is more costly as more work has been done on it, the program of quality control and tests should be graduated so that initially stringent tolerances and tests are gradually relaxed toward the required values as the operating conditions are approached.

5. The capabilities of the device by the time it reaches the opera-

tional state are fairly well known and prescribed. Most reliable results will be obtained by ensuring that the operator is well informed of the operating features of the device and that he observes instructions properly.

6. An active and intelligent maintenance and service program can accomplish much to improve reliability. The initial design must be formulated with the thought that preventive maintenance will be required and that service assessibility is a necessary design consideration.

7.2 Reliability Arithmetic

Our early mathematics training stresses the idea that an answer to a numerical problem is right or wrong. Furthermore, the mathematics involved when things are "probably right" or "probably wrong" becomes quite complicated, and most of us would prefer not to spend too much time on something that is so speculative.

However, we all know that in actual life the mathematical preciseness of the elementary classroom is somewhat impaired by such inexact things as weather forecasts, meetings not occurring exactly as scheduled, and things just not being made in the way specified. Hence the use of "the probability of a device or system performing" as part of the reliability definition is an acknowledgment of the realism of the work on reliability.

Although the derivation of many of the probability formulas is a strenuous exercise in mathematics, fortunately the day-by-day use of these formulas turns out to be little more complicated than the simple mathematical processes of elementary arithmetic. With the aid of charts and tables, the calculations required can be quite simple.[38]

Representative of some of the basic reliability formulas are the following:

1. *Constant Failure Rate.* For a device with a constant failure rate λ with time, the reliability, R, is $R = e^{-\lambda t}$, where e is the base of the natural logarithms, and t is time.

2. *Series Failure.* For a device with a number of independent parts, failure of any one of which causes the device to fail, R_T, the reliability of the device

$$R_T = R_1 \times R_2 \times R_3 \times R_4 \times R_n$$

where R_1, R_2, R_3, etc., are the reliability figures for the independent parts.

3. *Overall Failure Rate.* For a device such as described by 2 above in which the constant failure rates for the individual parts are λ_1, λ_2, λ_3, etc., the overall failure rate of the device, λ_T, is

$$\lambda_T = \lambda_1 + \lambda_2 + \lambda_3 + \cdots + \lambda_n$$

4. *Parallel Failure (Redundancy).* For a system consisting of two devices, one with a reliability R_1 and the other with a reliability R_2, either of which is capable of performing the desired function, the reliability of the system operating, R_T, is

$$R_T = R_1 + R_2 - R_1 R_2$$

Sample Reliability Problem

Whether one is concerned with supplying military systems as a prime contractor or subcontractor, or whether one merely supplies components to others who are contributing devices or assemblies for inclusion in military systems, the problem of reliability is going to have a more profound effect on his activities in the future. The following excerpts on the subject of reliability from an article in *Aviation Age* provide proof of this statement.

"Reliability requirements are being firmed up by ARDC. Some are being included in all new development contracts. According to Lt. Col. J. S. Lambert of ARDC's Aero-Electronics Directorate, U.S.A.F. will soon expect contractors to

(1) have reliability groups,
(2) indoctrinate personnel on the importance of reliability,
(3) monitor the quality of subcontractor's parts,
(4) be able to predict reliability of equipment,
(5) be able to test for reliability,
(6) prove the correlation between test and predictions,
(7) make necessary design corrections,
(8) provide operating and maintenance instructions to enhance reliability."

"USAF will require contractor to report within 90 days after receipt of contract on how he intends to meet reliability requirements."

"Reliability monitoring points will be set up at the following stages: preprototype, prototype, preproduction, demonstration of service readiness, full-scale production, and product improvement."

As an example of the way in which such reliability requirements

Reliability Arithmetic

may be presented, consider the following excerpt from a recent proposal request for an aircraft electrical generating system.

"The Primary Electrical Generating System shall have an in-flight mission reliability of 0.9997. This can be fulfilled by achieving a reliability of 0.982 for each of the two subsystems and designing so that no single failure will fail both of the two subsystems. A reliability of 0.982 corresponds to a mean time to failure of 250 hours. Tests shall be conducted that will establish the minimum reliability within a 90 per cent confidence level. Reliability is defined as the probability that the equipment will perform its required functions with no component failures that will cause the performance to deviate from the limits specified in the applicable performance specifications. This reliability factor shall be achieved within the environment utilization and mission requirements described herein."

Let us apply some of the probability formulas given above to understand better some of the reliability requirements listed. First, we will verify that the overall reliability of 0.9997 can be achieved with two subsystems each of 0.982 reliability operating in parallel.

A.
$$R_T = R_1 + R_2 - R_1 R_2$$
$$= 0.982 + 0.982 - 0.9643$$
$$R_T = 0.9997$$

B. Since $R = e^{-\lambda t}$, we can determine what the mission operating time will be for a mean time to failure of 250 hours. The mean time to failure, T, corresponds to a failure rate

$$\lambda = \frac{1}{T} = 0.004$$

Expressing $e^{-\lambda t}$ as the series expansion,

$$e^{-\lambda t} \simeq 1 - \lambda t + \frac{(\lambda t)^2}{2} - \cdots$$

$$0.982 \simeq 1 - 0.004 t$$

$$t = \frac{0.018}{0.004} = 4.5 \text{ hours per mission}$$

C. Each subsystem of the electrical generating system is itself composed of four major subassemblies. Assuming each of these to be capable of the same inherent reliability, what reliability requirement

Table 7.2-1. *Equipment Reliability*

Unit Minutes of Testing per Operational Minute	95% Confidence							90% Confidence					
	0 failures	1 failures	2 failures	3 failures	4 failures	5 failures	0 failures	1 failures	2 failures	3 failures	4 failures	5 failures
1	.05	.01	.002	.0004	.0001	.00005	.10	.02	.005	.001	.0003	.0001
2	.22	.09	.04	.02	.01	.005	.32	.14	.07	.04	.02	.009
3	.37	.20	.12	.08	.05	.03	.47	.28	.17	.11	.07	.05
4	.47	.30	.21	.15	.10	.07	.56	.37	.25	.19	.13	.09
5	.55	.38	.28	.21	.16	.12	.63	.46	.35	.26	.20	.15
6	.61	.45	.35	.28	.22	.17	.68	.52	.42	.33	.27	.21
7	.65	.50	.41	.33	.27	.22	.72	.57	.47	.38	.32	.26
8	.68	.54	.45	.38	.31	.26	.75	.61	.51	.43	.36	.30
9	.72	.59	.50	.43	.36	.31	.77	.64	.55	.47	.41	.35
10	.74	.62	.53	.46	.40	.35	.79	.68	.59	.51	.45	.39
20	.86	.79	.73	.68	.63	.59	.89	.82	.77	.72	.67	.62
30	.900	.85	.81	.77	.73	.70	.924	.88	.84	.80	.77	.73
40	.925	.89	.85	.83	.79	.77	.942	.90	.88	.84	.82	.79
50	.940	.90	.88	.86	.83	.81	.954	.922	.90	.88	.85	.83
60	.950	.920	.90	.88	.86	.84	.962	.935	.913	.90	.88	.86
70	.957	.931	.910	.90	.88	.86	.967	.944	.924	.910	.89	.87
80	.962	.939	.920	.90	.89	.88	.971	.951	.933	.916	.90	.89
90	.967	.947	.931	.915	.90	.89	.974	.956	.940	.925	.910	.90

n												
100	.970	.952	.947	.923	.908	.900	.977	.961	.947	.933	.920	.906
200	.985	.976	.968	.961	.954	.948	.988	.980	.974	.967	.960	.953
300	.990	.984	.979	.974	.969	.965	.9924	.987	.982	.980	.974	.969
400	.9925	.988	.984	.981	.977	.974	.9942	.9901	.987	.983	.980	.976
500	.9940	.9904	.987	.985	.982	.979	.9954	.9922	.989	.987	.984	.981
600	.9950	.9920	.990	.987	.985	.982	.9962	.9935	.9913	.989	.987	.984
700	.9957	.9931	.9910	.989	.987	.985	.9967	.9944	.9924	.9904	.988	.986
800	.9962	.9939	.9920	.990	.988	.987	.9971	.9951	.9933	.9916	.990	.988
900	.9967	.9957	.9931	.9915	.989	.988	.9974	.9956	.9940	.9925	.9910	.989
1,000	.9970	.9952	.9947	.9923	.9908	.9900	.9977	.9961	.9947	.9933	.9920	.9906
2,000	.9988	.9976	.9968	.9961	.9954	.9948	.9988	.9980	.9974	.9967	.9960	.9953
3,000	.9990	.9984	.9979	.9974	.9969	.9965	.9992	.9987	.9982	.9980	.9974	.9969
4,000	.9993	.9988	.9984	.9981	.9977	.9974	.9994	.9990	.9987	.9983	.9980	.9976
5,000	.9994	.9990	.9987	.9985	.9982	.9979	.9995	.9992	.9989	.9987	.9984	.9981
6,000	.9995	.9992	.9990	.9987	.9985	.9982	.9996	.9994	.9991	.9989	.9987	.9984
7,000	.9996	.9993	.9991	.9989	.9987	.9985	.9997	.9994	.9992	.9990	.9988	.9986
8,000	.9996	.9994	.9992	.9990	.9988	.9987	.9997	.9995	.9993	.9992	.9990	.9988
9,000	.9997	.9995	.9993	.9991	.9989	.9988	.9997	.9996	.9994	.9993	.9991	.9989
10,000	.99970	.99952	.99947	.99923	.99908	.99900	.99977	.99961	.99947	.99933	.99920	.99906
20,000	.99985	.99926	.99968	.99961	.99954	.99948	.99988	.99980	.99974	.99967	.99960	.99953
30,000	.99990	.99984	.99979	.99974	.99969	.99965	.99992	.99987	.99982	.99980	.99974	.99969
40,000	.99993	.99988	.99984	.99981	.99977	.99974	.99994	.99990	.99987	.99983	.99980	.99976
50,000	.99994	.99990	.99987	.99985	.99982	.99979	.99995	.99992	.99989	.99987	.99984	.99981
60,000	.99995	.99992	.99990	.99987	.99985	.99982	.99996	.99994	.99991	.99989	.99987	.99984
70,000	.99996	.99993	.99991	.99989	.99987	.99985	.99997	.99994	.99992	.99990	.99988	.99986
80,000	.99996	.99994	.99992	.99990	.99988	.99987	.99997	.99995	.99993	.99992	.99990	.99988
90,000	.99997	.99995	.99993	.99991	.99989	.99988	.99997	.99996	.99994	.99993	.99991	.99989
100,000	.99997	.99995	.99994	.99992	.99991	.99990	.99998	.99996	.99995	.99993	.99992	.99991

should each have so that the subsystem reliability will be 0.982?

$$R_T = R_1 \times R_2 \times R_3 \times R_4$$
$$0.982 = R_1{}^4$$

Letting $(1 - a) = R_1$,

$$0.982 = (1 - a)^4 \cong 1 - 4a$$

$$a = \frac{0.018}{4} = 0.0045$$

and
$R_1 = 0.9955$ for each major subassembly
  Viewed in another fashion,

$$\lambda = \lambda_1 + \lambda_2 + \lambda_3 + \lambda_4$$

$$\frac{1}{250} = 4\left(\frac{1}{T_1}\right)$$

or $T_1 = 1000$ hours mean time to failure for each major subassembly
  D. To establish the 90 per cent confidence level desired, one may refer to an equipment reliability table such as shown in Table 7.2-1

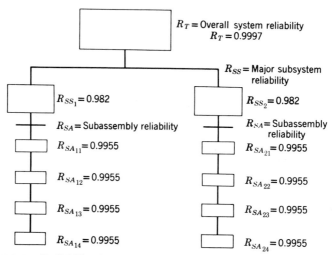

Figure 7.2-1. Reliability block diagram showing requirements at system, subsystem, and subassembly levels.

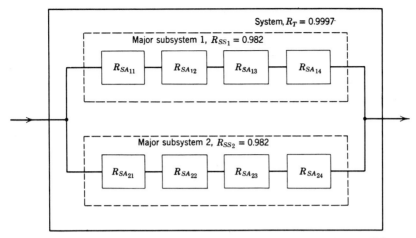

Figure 7.2-2. Alternative form of representing system reliability at different levels.

and determine the number of unit minutes of testing per operational minute.

For the subsystem reliability of 0.982 with 90 per cent confidence the table reveals that 300 unit minutes per operational minute with two failures yield a reliability of 0.982. The operational time is in reality 4.5 hours so that $4.5 \times 300 = 1350$ subsystem hours with two failures allowed will provide the confidence level sought. Depending on the cost and time one is willing to accept for the desired confidence, the number of subsystems and the time of testing will be altered accordingly to meet the 1350 subsystem hours. Using fewer failures as an acceptable number, a shorter testing time would be required or a fewer number of subsystems could be used with an attendant reduction in time and/or cost. Also, tests on smaller subassemblies may be considered appropriate in addition to complete subsystem tests, with the extent of testing being that corresponding to the required reliability figure for the subassembly or unit involved.

The preceding material has served to indicate the nature of the reliability problem in a concrete and quantitative form. It will serve to focus attention to some of the various areas of reliability to which the following section will be devoted. Figure 7.2-1 is one form of reliability block diagram showing the requirements at the system, subsystem, and subassembly levels. Figure 7.2-2 illustrates another way in which the reliability of the system at different levels can be indicated.

7.3 Designing for Reliability

Reliability has been defined as "the probability of a device or system performing adequately for the period of time intended under the operating conditions encountered."[5] The designer must, therefore, know the performance characteristics of his device or system and its performance requirements. He must know the period of time intended for this performance as well as the operating conditions encountered. The designer must meet a number of different performance requirements so that his equipment will perform satisfactorily under a large variety of external environmental conditions. Good design procedure must be used to meet these requirements if this is physically and economically possible. If it is not possible, the environment must be modified or other means must be established whereby the requirements can be met.

A generalized design approach that has been found to be successful in many cases consists of the following major steps:

A. Establish stress requirements.
B. Arrive at tentative designs.
C. Estimate failure rate.
D. Identify failure modes and consequences.
E. Establish tolerances and tests.
F. Review design and feedback results.

A. Establish stress requirements. The first problem that the designer faces is that of determining what stresses will be encountered by his device under the conditions of operation. Although these stress requirements may be provided in the specifications with which the designer works, frequently the requirements are based on a certain assumed condition of operation for the system of which this particular device is a part. To the extent that the stress requirements, such as temperature, vibration, humidity, moisture, shock, and other conditions, can be met with a reasonable design, the necessity for questioning the environmental specifications is not significant. However, when these requirements impose an unreasonable burden on the device, it is worthwhile to explore further the system of which the device is a part to consider other methods of achieving the end result than the ones initially contemplated.

In addition to the stress requirements as far as environment is concerned, it is also of the greatest significance for the designer to determine the performance needs for his device. What is the perfor-

Designing for Reliability

mance and what are the tolerances allowed on this performance? If the needs are for certain maximum power outputs, the question of how long these outputs must be available should be established initially. In addition, the degree of variation which will be allowed, i.e., the tolerances on performance, should likewise be established early in the design. Furthermore, the characteristics of the inputs from other devices should be known at the outset. These include not only average or nominal conditions but also transient, resonant, and other dynamic phenomena which may alter appreciably the actual input or environmental conditions that the device will encounter.

Although it is difficult for the designer to understand all the ramifications of the system in which the device he is working on must operate, nevertheless, to the extent that he is able to achieve this understanding, he can do a better job of meeting the reliability requirements of his device.

B. Arrive at tentative designs. To the extent that the device being designed has already been built and is operating successfully, it is desirable that the device for a given application be one which can be taken from existing designs. Not only can better test and field information be obtained on such components, but also the manufacturing procedures are established and the whole process of education for reliability may be already in operation.

Where new designs are necessary, it is important that the standards and practices established in the past as being suitable for most reliable performance be used to the greatest extent possible. Since in many cases these standards and practices have been built on successful past performance, they stand the greatest chance of satisfactory operation in the future. Furthermore, increasing effort is being made in the preparation of standards and practices to include reliability information as such, since these standards can provide a basis for designs of greater inherent reliability.

A major principle for the designer to keep in mind for reliability is use of conservatively rated components. This may mean using components at one-half their rating or less so that ample reserve capacity is available to meet performance demands not originally planned for. Also, in this way variations in manufacturing tolerances should not so degrade the capability of the device or component that it is unable to meet the performance requirements. In arriving at conservatively rated components, the designer should keep in mind the overall effect of the selection. For example, if components that all are excessive in weight are used in order to meet a conservative strength design, the equipment may "never get off the ground." There must be a good design balance as well as conservatively rated components.

C. Estimate failure rate. Having established the stress requirements and arrived at a tentative design, it is desirable to test this design to determine its ability to meet the desired performance requirements. There are a number of ways of estimating this failure rate, and three will be mentioned here.

The first is on a statistical random failure basis. This is particularly applicable to electrical or electronic designs having a large number of components for each of which there may be data in regard to the expected failure rate. For example, a summary of the number of tubes of different types, the number of resistances operating at certain percentages of their power rating, the number of capacitors, diodes, transistors, transformers, reactors, motors, and other devices can be tabulated, and the failure rates associated with the particular operation of each of these devices in the design can be established. This information can yield a numerical value for the estimated failure rate of the device on the basis of random failures, random failures being the cause of unsatisfactory performance. Depending on whether or not this figure is satisfactory, the designer may have to review his design with the thought of improving the failure rate figure to achieve adequate reliability. By providing duplicate equipments in part or in total, it may be possible to improve significantly the reliability estimated for the system.

The second method for establishing failure rates is that of considering the strength-stress relationship for the device as a function of its life. Figure 7.3-1 shows the frequency of occurrence of the strength and of the stress encountered for a representative device. In such

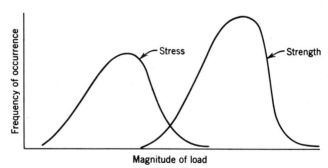

Figure 7.3-1. Generalized stress and strength frequency of occurrence versus magnitude of load for a representative device.

a generalized presentation of the stress requirement, it is important to take into account unusual or transient phenomena, for example, shocks encountered during transportation or shipment, or transients caused by voltage switching or other overvoltage phenomena, or overstressing due to temporary thermal overloads. Although it is not possible to pinpoint when these unusual conditions are likely to be encountered, the ability of the system or device to survive stresses of this kind must be analyzed.

The third method for establishing failure rates is the experimental one of testing. Breadboard or laboratory models can be constructed to demonstrate the significant factors of the design, and these models can be tested in the laboratory under environmental conditions that are related to those to be encountered in practice. In general, the goal of such laboratory tests will be to meet standards sufficiently higher than will be encountered in the actual operational environment so as to provide assurances that normal loads or stresses can be met without failure.

D. Identify failure modes and consequences. The proceeding three steps have served to make the designer familiar with the normal method of operation of the equipment as well as the environment, both normal and abnormal, in which it must work and the characteristics of the rest of the system of which it is a part. It is now essential that the designer review the equipment in painstaking detail to determine in what myriad of ways the equipment parts can fail to operate properly. Strength, manufacturing tolerances, assembly, environmental and operational wear, and other types of problems must be considered.

Accompanying this identification of the failure modes, the designers must determine the consequences of failure. If the particular part or equipment experiences each of the failures possible, what will be the effects on the desired output of this or associated equipment? The failure consequences provide a basis for weighting the importance of altering the design to minimize the possibility of each sort of failure.

Although the identification of failure modes and consequences is by no means a precise discipline, an organized method of failure analysis can prove beneficial in reducing the probability of equipment failure.

E. Establish tolerances and tests. The reliability definition specifically states that the device must perform adequately for the period of time intended. Therefore, it is essential that the design be so chosen that the device will perform satisfactorily over its entire life. In other words, the device should be "design centered." Not only should its performance fall within the range of acceptable values initially, but also, in spite of aging and other changes which take place during the life of the device, it must continue to perform satisfactorily. The design centering must take into account the effect of tolerances in the initial values of the parameters. Although nominal values for the various components making up the device must be listed, the values for their tolerances should likewise be included. In addition to the tolerances which exist because of variation in the manufacture of the different components, parameter changes which take place as a function of changing environment must also receive proper attention. For an equipment which must operate over a range of $-60°C$ to $+200°C$, it is obvious that such features as resistance and mechanical and physical dimensions, as well as many other characteristics, will change. The performance of the device must be determined with these changed parameters as well as with the normal parameters.

Equations should be written in terms of the variation in performance which will be encountered as a result of changes in the different parameters. On this basis it will be possible to combine the effects of different parameter variations to arrive at a suitable overall estimate of the change in the total performance of the device.

Statistical methods can be employed to advantage to determine the probability of variations in the parameters of the amounts indicated. When these parameter changes occur in a random fashion, this statistical technique can be used. However, when the cause for variation in a parameter is definitely established, as, for example, the fact that a certain resistance has a particular temperature coefficient or a certain capacitance has another temperature coefficient, the effect of changes in different parameters may be correlated, and therefore it is essential that the effect of changes in parameters be coupled with the physical reason for these changes.

Having arrived at a good understanding of the effect of parameter changes on performance and having determined the effect of aging on performance, it is necessary to establish a graduated set of tests so that there can be a continual reduction in severity as more and more work is done on an assembly or a subassembly.

F. Review design and feed back results from factory and from field. At the "completion" of the formal design and with its reduction to a production basis, it is essential that a design review be held at which time representatives from all phases of engineering, manufacturing, quality control, and reliability be present to consider all aspects of the "final" equipment design. Specialists in the various skills and technologies involved should be brought in to contribute from their experience with similar devices.

Although the design as initially established on the basis of the known stress requirements and performance needs may prove to be adequate for the purpose intended, it is essential that as part of the design procedure methods be set up for obtaining the results of preliminary and production tests from the factory as well as of experience from the field. On this basis it will be possible to check the original design assumptions as to environment and performance; it will also be possible to determine whether the tolerances and design requirements established initially are adequate or perhaps excessively stringent. By maintaining such reviews in the light of the factory and field experience, improved designs with either greater reliability or lower cost with adequate reliability can be obtained. As long as this review and feedback procedure is incorporated into the initial design planning, its chances of success are considerably greater and the cost involved may be much less than would be required by some last-minute inclusion of this type of program.

Stress Requirements Including Environment

The reliability of a device is markedly affected by its ability to perform adequately "under the operating conditions encountered." Since the operating conditions encountered are many and varied, extending as they do over the entire life of the device or equipment from the time the materials going into it are made until the time it ceases to be used, the designer must understand all these conditions at the outset. Not only the magnitude of such environmental conditions as temperature, shock, vibration, voltage, pressure, and humidity, but also the rate of change, their interdependence upon one another, and their duration or time history are important. For this reason, it is essential that the designer have as complete a picture as possible of the functional factors influencing the environment and the performance requirements of the device.

Figure 7.1-4 has shown in block diagram form some of the functional factors which influence reliability over the life of the device, including its birth and origin. Roughly these areas of environment may be grouped into two major categories: the preoperational environment and the operation and maintenance environment.

Preoperational Environment. Once the device has been assembled it is necessary to quality-control and/or test it to be sure that it meets the performance and other conditions established by design or by contract. Although these preoperational environments are generally not of an exceptionally long duration, they do represent a stress on the device which in some cases is in excess of the design requirements and hence may influence the strength or inherent capability of the device to resist further stresses in the future. Since the designer usually helps to establish the quality control and test requirements, this aspect of the preoperational environment is reasonably well under his cognizance in most cases.

The shipment of a device, and its subsequent storage before installation and use, may represent one of the most drastic sets of environmental conditions that the device will encounter during its whole life. When packaged and shipped via regular commercial carriers or perhaps via rough military transport in foreign fields, the shock and humidity environment to which many devices have been subjected far exceeds their designers' wildest expectation, based solely on operational environment requirements. As a result, in many instances a significant portion of the equipment arriving for installation and operation has in times past been wholly inadequate for the needs of the job. Improved methods of shipment or packaging and storage have been developed and are currently in use. In a recent example of a condition in which storage materially reduced the reliability of the design, a transistorized equipment packaged in a cardboard shipping container was compromised because it was left out in a yard that was flooded during the spring thaw. As a result of this treatment during storage, the reliability of the transistorized equipment was severely criticized.

The installation phase of the preoperational environment is another area where testing and severe handling may impose environmental stresses on a device far more strenuous than would be encountered under normal conditions of operation.

Operation and Maintenance Environment. The environmental conditions to be encountered by a device under operation are of major concern to the designer. Certainly these conditions are the ones which most uniquely characterize the particular device and represent a starting point to which the initial strength capability must be directed. Frequently it is under the operational conditions that unusual temperature, electrical stresses, or mechanical shock and vibration are encountered. The presence of other allied equipment generating heat or producing unusual mechanical or electrical conditions contributes to the overall environment encountered by the device during its operation.

Designing for Reliability

Maintenance and service represent an important influence on the operational performance of a device in that they can provide a pre-operational environment of their own. For example, the capability of equipment re-entering a new operational phase is appreciably influenced by the preventive maintenance and service treatment that it has just received. Obviously some devices are such that they are destroyed in operation, and the opportunity for maintenance and servicing is present only during the preoperational phase. However, for such equipment as airplane engines, where extensive use of regular maintenance and service procedures has been able to greatly improve operational capability, it is quite apparent that the maintenance and service factor is of great importance.

Environmental Factors to be Considered. Although the nature of the environmental factors to be encountered will vary considerably with the type of device being designed and the application to which it will be put, certain major factors can be singled out. These are as follows: electrical, mechanical, transient, and frequency response. In the electrical area such considerations as signal levels obtainable, noise inputs, power supplies, miscellaneous inductive coupling signals, voltage levels, voltage gradients, and similar factors must be included. In the mechanical environment such factors as shock and vibrations, mounting fatigue, temperature, pressure, humidity, sand, and dust are significant.

Under the transient or dynamic conditions of environment, it is important that the time characteristics of these various factors, whether they be electrical or mechanical, be clearly understood and stated. For example, if a heating cycle is of the nature illustrated in Figure 7.3-2, it is of significance to indicate the time relationship shown rather than merely to specify that the average value of tem-

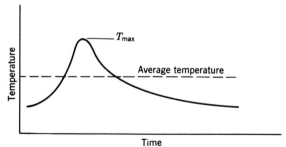

Figure 7.3-2. Temperature-time environment for a device.

perature has the magnitude indicated, as, for example, by the dotted line.

In terms of electrical transients, steep wave fronts may cause abrupt rates of change of currents and accompanying high voltages in closely coupled circuits. Furthermore, since the electrical life of a device or component is generally sharply dependent on the maximum value of the positive or negative voltages, it is very important that sharp voltage peaks such as those shown by E_p and $-E_v$ on Figure 7.3-3 be avoided. Although the effect of these peaks is negligible on the average voltage E_{avg}, their effect on capacitors and transistors may be fatal. Unusual effects generally occur at the instant of switching, and for this reason devices such as relays or switches require particular attention. Capacitors or nonlinear devices such as thyrite or diodes may be used effectively to limit the high-voltage spikes that would otherwise be so damaging at switching times.

In regard to frequency response, this characteristic applies both to mechanical and electrical devices. Figure 7.3-4 shows a plot of amplitude versus frequency and indicates two conditions. The one shown by a solid line has a high resonant peak; the other, indicated by the dotted line, has a much lower resonance condition. For purposes of the nominal operation of the device it might appear that the environment was such that the frequencies need not exceed that shown by f_1 on Figure 7.3-4. However, in practice, for various reasons it often occurs that inputs of frequencies considerably greater are encountered. For this reason, knowledge of the overall frequency

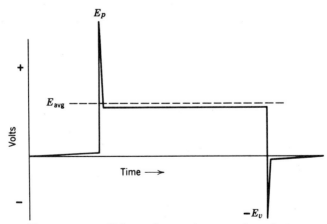

Figure 7.3-3. Voltage-time environment for a device.

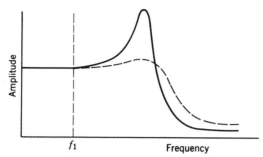

Figure 7.3-4. Typical frequency-responses for a device.

response characteristic shown in Figure 7.3-4 is extremely important. For applications having low frequency requirements, a device with the characteristics shown by the dotted frequency response is more desirable than the one represented by the solid line.

Establish Preliminary Design

In establishing a preliminary design, the designer is aided considerably by having available a functional model to represent the life of the device. Figure 7.3-5 shows the different phases of the life of a device. It serves to focus the attention of the designer on all the different environments that the equipment will face. By means of such a model it is possible for the designer to set performance levels at different stages of the process so that a graduated series of levels can be established. Looking at the overall design, it is possible to arrange for suitable subunits and an overall systems arrangement as far as voltage levels and other factors are concerned. By keeping in mind performance levels at such different phases as research and development, preliminary design, production design, production, and quality control and test, the designer, during the research, development, and preliminary design, can gather data for use in these later phases of the life of the equipment.

An important phase of the preliminary design is that of establishing firmly specifications compatible with the design results that are obtainable. The specifications for the equipment should be firmly established during this preliminary design period if they have not already been set forth.

In Figure 7.3-5, one notes a number of dotted lines indicating feedback of information at different stages of the model. The *design review* consists of a thorough and critical review of the equipment

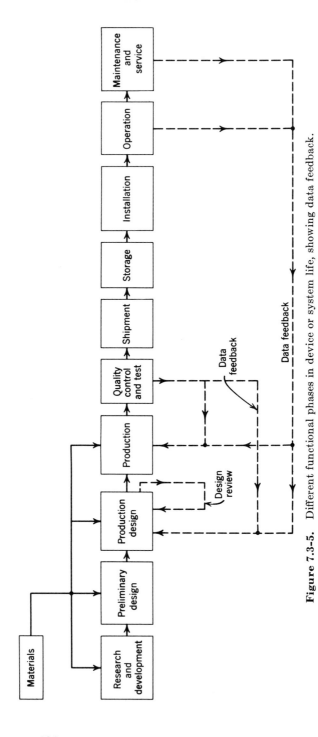

Figure 7.3-5. Different functional phases in device or system life, showing data feedback.

294

Designing for Reliability

design, at the completion of the drawings and/or of a prototype unit, by skilled designers or experts in the fields in addition to those involved with the equipment. Manufacturing, service, reliability, and systems personnel should also be included in this design review. *Data feedback* from quality control and test to the interested design and manufacturing groups is valuable. Likewise operation, maintenance, and service data quickly sent back to design and manufacture groups should help to make more reliable equipment available sooner. These feedback functions, by providing information earlier in the life of the device, will serve to reconcile the design with the facts and bring about more quickly a reliable equipment.

A primary precept that is invaluable in providing reliable design is the *derating of components to provide an adequate factor of safety*. Frequently, standard practices are now recommending use of such components as resistors, capacitors, and tubes at approximately 50

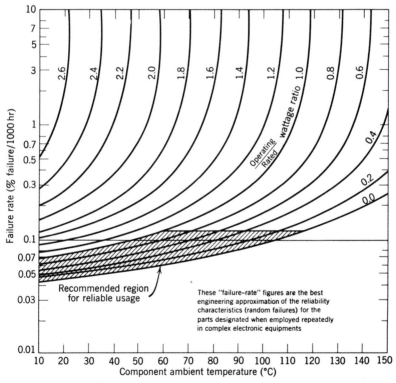

Figure 7.3-6. Predicted failure rates for composition resistors (MIL-R-IIA characteristic GF).[7]

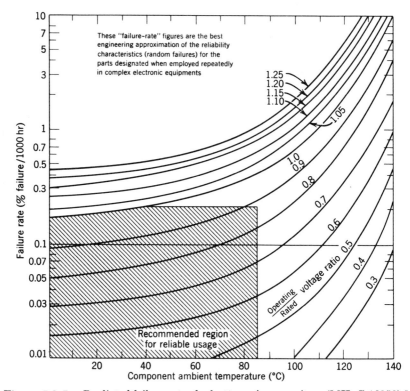

Figure 7.3-7. Predicted failure rates for button mica capacitors (MIL-C-10950).[7]

per cent of their rated values, to obtain markedly increased life. Figures 7.3-6 and 7.3-7 show how reduced failure rates result from operation at low wattage and voltage ratios for resistors and capacitors respectively.[7]

Use of Existing Designs. In the case of electronics or transistor equipment there has been a growing trend in recent years within industry and government to make use of so-called standard or preferred circuits in design. Much of the activity in this field has been carried on at the National Bureau of Standards for the Bureau of Aeronautics Department for the Navy and is described in *Handbook of Preferred Circuits,* Navy Aeronautical Electronic Equipment.[6]

Some of the important advantages inherent in the use of preferred circuits are as follows:

1. The time and cost involved in circuit design are reduced.
2. The general overall quality of the design is improved since the

Designing for Reliability 297

preferred circuits can be selected on a basis of reliability and optimum performance.

3. The designer can have available more precise information on the performance of the more common circuits which he may employ in his system and is thus able to tailor the system to avoid unrealistic specifications for the circuit and components.

4. Preferred circuits will help reduce manufacturing costs and simplify repair and maintenance problems.

5. In advanced circuit design, preferred circuits will provide a good basis for the evaluation and specifications of the new design.

In other fields as well as electrical and electronic equipment it is apparent that standard designs, where they are applicable, can do much to enhance system reliability.

Example: Transistor Circuits

As an indication of an approach that has been found useful in the standardization of transistor circuit design, a compilation has been made of "transistor preferred circuits" from 295 reports. These circuits were chosen primarily on the basis of general applicability, but consideration also was given to such factors as stability, operation with production-type transistors, operation over wide temperature ranges, and availability of adequate design and performance data. For circuits which are similar in design and application only the most promising circuit is listed, and the sources of the other circuits are given in the references.

It should be emphasized that these circuits are not suggested as preferred circuits in either their function or their design, but rather should serve as a general background from which future transistor preferred circuits can be derived.

Table 7.3-1 illustrates the indexing system employed and shows the major categories of the circuits for which data are available. Table 7.3-2 shows a flip-flop circuit and shift register connection contained in this report of transistor preferred circuits. Figure 7.3-8 indicates the circuit diagram contained in this report and accompanying the information in Table 7.3-2.

Example: Electronic Circuits

In regard to electronic circuits, the following quotation is from the foreword of *Handbook of Preferred Circuits*, Navy Aeronautical Electronic Equipment NAVAER 16-1-519:[6]

"The purpose of the preferred circuits manual is to encourage better engineering practice in the design of circuits for military electronic

Table 7.3-1. *Index of Transistor Circuits Used in Report "Transistor Preferred Circuits"*

1. Amplifiers
   A. Audio
   B. Pulse
   C. IF and r-f
   D. Wideband or video
2. Generators
   A. Pulse
   B. Sine wave
   C. Square wave
   D. Saw tooth
3. Computer Applications
   A. Flip-flops
   B. Logic
   C. Core drivers
4. Power Supplies
   A. Voltage regulators
   B. High voltage
   C. A-c/d-c, d-c/a-c converters
5. Receiver Circuits
   A. Mixers
   B. Detectors
   C. Automatic gain control
   D. Limiters
6. Miscellaneous Applications and Special
   A. Active filters

Table 7.3-2. *Flip-Flop Circuit and Shift Register Connection*

1. Source
   "Design of Non-saturating Transistor Flip-Flops"
2. Circuit Diagram
   Basic shift-register stage (500 kc). (See Figure 7.3-8.)
3. Performance Figures and General Information
   Temperature range 20–85°C. Loaded with 620 ohms, rise time = 0.8 μsec; fall time = 0.25 μsec. Circuits of this configuration will operate up to 2.5 Mc. Minimum trigger pulse at 500 kc, 5 volt 5 ma pulse. Standby power drain 224 mw.
4. Applications
   Digital computer as part of shift register circuit.
5. Design Considerations
   Allowable power supply variation is ± 50 per cent. This particular circuit operates at 500 kc. For complete design example see Appendix C of the source report.
6. Curves
   None

equipment. The manual may be considered as a compilation of good design practice, not toward the limiting of effort in circuit design, but as a standard against which true progress in circuit design may be measured.

"In the past, electronics engineering practice has been unique in the complete freedom of the circuit designer to choose his individual circuit elements with little regard to engineering already accomplished and proven. This lack of standardization was possible due to the availability of a vast choice of circuit elements and to the ease of correcting design errors in hand-assembled production. However, with the advent of mechanical assembly, mechanized production, and the prospect of automation electronics production, it is becoming more and more necessary for designs to be reliable and frozen prior to production and subsequent field use.

"Preliminary studies sponsored by the Bureau of Aeronautics indicate that a high percentage of circuit functions are subject to stan-

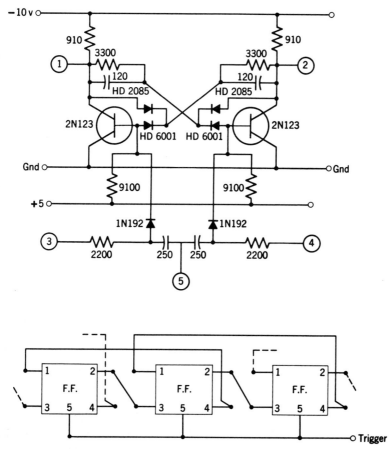

Figure 7.3-8. Circuit and interconnection diagrams for shift register.

dardization while still retaining the desired flexibility of equipment performance. For the military services such standardization will result in greater operational reliability, simpler maintenance training and procedures, shorter lead time on delivery of equipment, faster acceleration of production, more efficient use of engineers, lower original purchase prices of equipment, and fewer spare parts at field installations. For the equipment producer, standardization will result in lower production cost, quicker shifting of equipment from development to production phases, better conservation of scarce engineering manpower, and lower parts inventory."

Estimate Failure Rate

Having arrived at one or more tentative designs, it is necessary to make a preliminary estimation of the failure rate associated with this design. The most straightforward and simple method is to analyze the design on a statistical, random-failure basis. Determining the strength-stress relationship during the life of the equipment is a second method for estimating the failure rate. The third method is the tried-and-true use of experimental testing of prototype equipment.

Random-Failure Method. The random-failure method is based on the assumptions that all of the parts are operating in their normal useful life so that their failure rates are constant, and that the failure rates of the parts are in general so low that the system rate is simply the sum of the contributions of each of the individual elements.

The judicious selection of the suitable failure rate for the specific component at the selected failure mode becomes an all important part of the reliability estimation process. For the controls and accessories components, mortality data have been collected for certain relatively standard, well-developed parts. Table 7.3-3 is a sample of such a listing for designs of aircraft practice. The values have to be used with some caution and are primarily provided as guideposts. The data on electrical components in particular do not represent the best-quality parts. Judgment has to be applied in the use of the numbers to allow for special environmental and/or use factors, as well as special design considerations which depart from the nominal practice of the aircraft industry.

A commonly used first estimate for the failure rate of a device or circuit in the random-failure method is based on a count of the number and different kinds of elements and their corresponding failure

Table 7.3-3(a). *Element and Subassembly Failure-Rate Data to be Used for Estimation of Component Failure Rates*

Failure rate per million operating hours based upon not exceeding the prescribed overhaul life of the design

Part, Element, or Subassembly Name	Development		Operational	
	Early	Late	Maintenance Type Failure	Confirmed Operational Squawk Type Failure
Predominantly Hydromechanical				
Adjustment	30	15		2
Bearings				
Rotary sleeve shaft	200	120	40	10
Translatory sleeve shaft	42	25	20	1
Rotary roller bearing	5		1	0.5
Bellows, null type	250	140	5	2
Motor in excess of				
0.05 in. stroke	300	200	20	8
Capillary tubes, external	800	50	50	10
Internal	500	20	10	2
Casting or housing	30	10	2	1
Cam and follower assembly	500	100	15	10
Connections, fuel			25	2
Lube			20	1
Hydraulic			30	3
Quick disconnect			80	5
Coupling, for Lever to Device				
Rotary to translatory	500	10	10	1
Rotary to rotary	50	5	1	0.01
Fittings (*see* Connections)				
Filter (self washing)			10	1
Flange and gasket				
Not subject to field service	50	30	1	0.2
Subject to field service	20	10	10	2
Flyweight assembly				
Bearing, shaft, and shoes	300	100	40	15
Gear assembly percontact				
Nonpumping	20	10	4	2
Lube pumping	250	100	85	45
Fuel pumping	50	25	20	10
Lines, flex hoses	40	20	30	2
Lever, link	5	...	...	0.01
Manifolds				3
Plus per weld				0.2
Mounting legs or bosses	50	5	2.5	0.25

Table 7.3-3(b). *Component Failure-Rate Data to be Used for Estimation of Failure Rates of Aircraft-Type Applications*

These are average or normal failure rates in parts per million operating hours with overhaul. It should be recognized that these may be improved with special techniques in design, development and use.

Components (Serialed Parts)	Valid Removal for Operational Squawk
Actuator	50
Accumulator	...
Alternator	225
Amplifier, electronic	
Temperature	800
Amplifier, magnetic	
Temperature	250
Control, hydraulic	
Fuel, afterburner	200
Fuel, main engine (less sensor CIT)	350
Nozzle area	200
Temperature sensor (part not included in MFC above)	40
Cable, electrical	1
Per lead connection	0.5
Cable, feedback plus sheave wheel	20
Filter, relieving	60
Nonrelieving	100
Heater, fuel	10
Cooler, fuel-oil	1
Fuel-air	10
Lines, hose	3
Rigid tube	10
Line, rigid plus bellows	30
Motors, hydraulic	175
Electrical	150
Pump, main fuel, gear	
Single element	60 (Leakage)
Double element	80 (Leakage)
Pump, fuel centrifugal	
Lube supply	200
Scavenge	350
Hydraulic, piston	250
Reservoir, lube or hydraulic	
engine mounted	50
Sensor, temperature T_2	30
Temperature T_5	...
Tachometer, amplidyne	125
Valve, check or relief	6

Table 7.3-3(b) (Continued)

Components (Serialed Parts)	Valid Removal for Operational Squawk
Predominantly Electrical	
Capacitors	0.9
Choke	3
Circuit breakers, thermal	25
Connectors	2–5
Diodes	10
Zener	30
Fuses	25
Generators	100
D-c	160
Tach.	150
Inverters	120–600
Magnetos	565
Motor (electric)	150
Potentiometers	100
Reactors	20
Rectifiers	2
Relays	300
Resistors, carbon	1
Wirewound	0.5
Rheostats	20
Solder connections	0.5
Solenoid	13
Starters	303
Switches	90
Micro	33
Toggle	7
Syncros	100
Terminals	1
Thermister	10
Thermocouples	400
Transducers	27
Transformers	0.1–50
Transistors	12
T-R units (Reg.)	200
Tube (all vacuum)	100
Variac	6
Voltage regulators	200

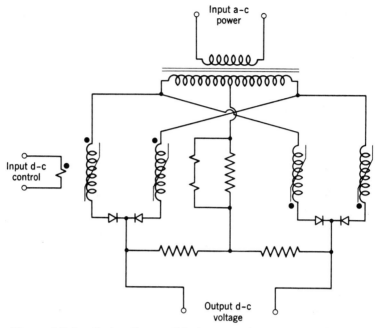

Figure 7.3-9. Push-pull reversible d-c saturating magnetic amplifier.

rates. Consider the following example for the push-pull reversible d-c magnetic amplifier shown in Figure 7.3-9.

Type of Component	Number, N	Failure Rate $\times 10^{-6}$	Failure Contribution, $N \times F \times 10^{-6}$
Diode	4	10	40
Transformer	1	3	3
Reactor (magnetic amplifier)	1	20	20
Resistor	3	1	3
Total			66

Complete magnetic amplifier failure rate per 10^6 hours = 66
Magnetic amplifier mean time to failure = $1/(66 \times 10^{-6})$
$= 15{,}000$ hours

Assuming a mission time of 3 hours,
Reliability $= e^{-t/M} = e^{-3/15{,}000} = 0.9998$

Although the above calculation is indeed simple and very approximate, it serves to provide a first estimate as to a reasonable value

Designing for Reliability

for the reliability of a magnetic amplifier for this short mission of 3 hours. This example shows the importance of both the basic failure-rate data and the mission duration in arriving at the reliability of a device or circuit. Obviously, more refined calculations are possible and at times are necessary. This method is simple and is often used to obtain a first estimate of the reliability of a proposed design.

A method of recognizing the variability of failure-rate data with design conditions is the use of the failure-rate modifier in the form

$$F/R = (F/R)_{\text{nom}} \times K_1 \times K_2 \times K_3 \times K_4 \qquad (7.3\text{-}1)$$

where $(F/R)_{\text{nom}}$ is the nominal failure rate,
K_1, K_2, K_3, K_4 are the failure-rate modifiers.

Examples of representative curves of such failure rate modifiers are shown in Figure 7.3-10 through 7.3-13 for a journal bearing and the spool valve and sleeve.

The failure rate, λ (failures per million hours), is related to the reliability, $R(t)$, the probability of success for a mission of time t, by the well-known exponential relationship for the case of random

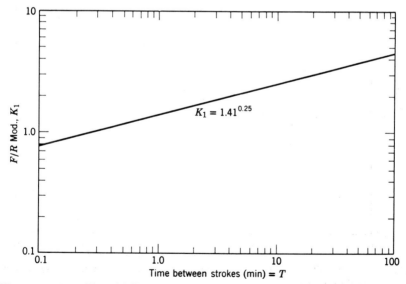

Figure 7.3-10. Plot of failure rate modifier K_1 for spool valve and sleeve seize-silt effects.

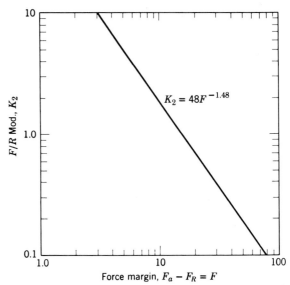

Figure 7.3-11. Plot of failure rate modifier K_2 for spool valve and sleeve silt effects.

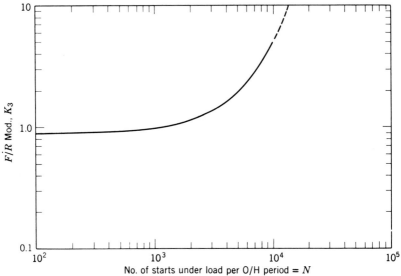

Figure 7.3-12. Plot of failure rate modifier K_3 for journal bearing, hydrodynamic wear, and abrasion effects.

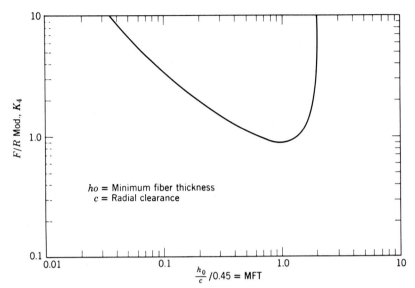

Figure 7.3-13. Plot of failure rate modifier K_4 for journal bearing, hydrodynamic wear, and abrasion effects.

single failures and low failure rates:

$$R(t) = e^{-\lambda t \times 10^6} \qquad (7.3\text{-}2)$$

or

$$R(t) = 1 - \lambda t \times 10^6 \qquad (7.3\text{-}3)$$

assuming that λ is by far the dominant failure rate. If the system consists of many components each of which could fail individually at rates of similar orders of magnitude, if no redundancies are used and any one of the failures could singly cause a system failure, if the degradation of one part does not affect the failure rate of another, and if all components are active for the time t, then the system failure rate is simply given by

$$\lambda_s = \sum_1^n \lambda_i \qquad (7.3\text{-}4)$$

Thus it becomes only a matter of establishing individual failure rates in order to compute a prediction of the system reliability.

Redundancy. To enhance the probability of a given function being performed, it is desirable that there be more than one equipment or method for accomplishing the function, any one of which alone is capable of doing the job. The application of alternative methods or equipments is called redundancy or the use of redundant equipments. Redundancy provides a real opportunity for ensuring the possibility of reliable operation when all means have been exhausted for the reliable design of one equipment by itself.

Where an event has a probability p of occurring, and it is desired to know the probability of similar events happening concurrently in n similar systems, all possible event combinations are given by the binomial expansion of $[p + (1-p)]$, e.g.,

$$[(1-p) + p]^n = (1-p)^n + n(1-p)^{n-1}p + \frac{n(n-1)}{2!}$$

$$\times (1-p)^{n-2}p^2 \ldots n(1-p)p^{n-1} + p^n = 1 \quad (7.3\text{-}5)$$

In this equation, the probability of exactly r events occurring concurrently in the n systems is given by the $(r+1)$th term of the expansion. For example, if one were to roll three dice together, the probability of getting exactly two 2's (no more, no less) would be given by the third term of Equation 7.3-5. For this case $n = 3$, $p = \frac{1}{6}$, and

$$\frac{n(n-1)}{2!}(1-p)^{n-2}p^2 = \frac{3(3-1)}{2}(1-\tfrac{1}{6})(\tfrac{1}{6})^2 = \tfrac{5}{72} \quad (7.3\text{-}6)$$

The coefficient of each term indicates the number of possible combinations fulfilling the problem requirements exactly; for example, in the above case $n(n-1)/2! = 3$ indicates that three combinations of events can occur that give exactly two 2's.

The assessment of the reliability enhancement of systems from internal redundancy is a closely analogous problem to that of the above example. In such cases the event probability is the failure probability of some unit in a system, for example, an engine in an airplane, and our interest lies in determining the probability of more than one engine failure in a multiengine plane which could cause loss of the airframe.

For this class of problem it is convenient to condense the information of Equation 7.3-5 into the following form by writing a general expression for the terms in the equation:

$$P = \frac{n!}{F!(n-F)!}(1-p)^{n-F}p^F$$
$$= C(1-p)^{n-F}p^F \quad (7.3\text{-}7)$$

Designing for Reliability

Table 7.3-4. Values of Coefficient C

No. of Failures, F	No. of Items, n					
	1	2	3	4	5	6
0	1	1	1	1	1	1
1	1	2	3	4	5	6
2	...	1	3	6	10	15
3	...	...	1	4	10	20
4	...	...	...	1	5	15
5	...	...	...	...	1	6
6	...	...	...	...	...	1

where p is the failure probability of one unit in the system,
  n is the number of units in the system,
  F is a particular number of failed units causing system failure,
  P is the probability of system failure.

Values of the constant C for values of n and F ordinarily encountered are given in Table 7.3-4.

Example 1. Consider a jet-engine nozzle actuator system depicted schematically in Figure 7.3-14.

The common hydraulic system supplies six linear hydraulic actuators, each of which acts on the adjustable nozzle mechanism

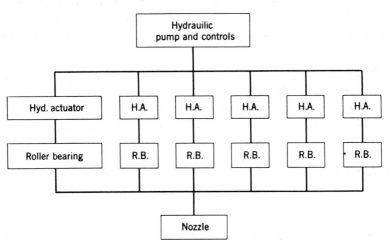

Figure 7.3-14. Redundant jet-engine nozzle actuator system shown schematically.

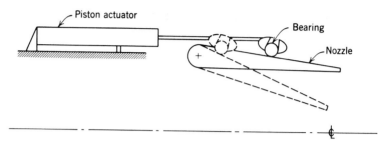

Figure 7.3-15. Piston actuator-bearing-nozzle arrangement.

through a roller-bearing-cam arrangement as shown in Figure 7.3-15. These six actuator-bearing combinations are equally spaced around the circumference of the nozzle.

The following facts are known:

1. The hydraulic actuator failures are piston seal types so that one failure does not freeze the piston and consequently the nozzle position.

2. In failure, the roller bearings in the assembly tend to freeze, which increases loading on the actuating system but does not, up to two nonadjacent failures, prevent operation with this increased load.

3. One, or at most two, nonadjacent failures are tolerable. That is, the design is such that a 33 per cent increased loading on the remaining four actuators is permissible, but binding of the nozzle mechanism would result if the two failing actuators were adjacent. Simultaneous failure of more than two actuator-bearing units would prevent nozzle operation.

From this information, we can estimate the probability of various failure modes as follows:

(a) *Probability of two adjacent failures.* Table 7.3-4 indicates that fifteen double-failure combinations are possible; however, some of these are nonadjacent. Inspection of the system design indicates that only six adjacent failure combinations are possible. Hence the probability of system failure due to such an event combination is

$$P_{2A} = 6p^2(1-p)^4$$

where p is the failure probability of any one actuator-bearing combination.

Designing for Reliability

(b) *Probability of three or more failures.* These probabilities are estimated from Equation 7.3-7 and Table 7.3-4 as follows:

$$P_3 = 20p^3(1-p)^3$$
$$P_4 = 15p^4(1-p)^2$$
$$P_5 = 6p^5(1-p)$$
$$P_6 = p^6$$
(7.3-8)

These relations can be combined to give the probability of nonfailure of the system, R, as follows:

$$R = 1 - 6p^2(1-p)^4 - 20p^3(1-\ )p^3 \\ - 15p^4(1-p)^2 - 6p^5(1-p) - p^6 \quad (7.3\text{-}9)$$

From data on the failure rates experienced with hydraulic actuators and bearings, the value of p for use in Equation 7.3-9 can be estimated by means of the addition rule, e.g.,

$$p = p_b + p_a - p_a p_b$$

where p_a is the probability of an actuator failure,

p_b is the probability of a bearing failure.

Equation 7.3-9 can then be used to estimate the reliability of the complete nozzle actuator system.

For the valid application of this type of analysis, it is essential that a requirement relating to interaction be observed. This requirement is that the redundant units be truly independent so that the failure of one does not affect the system or the duplicate unit in any way. To gain assurance on this point, it is often necessary to carry out a detailed failure analysis to determine the effect of different modes of failure of the component and/or the effectiveness of any means of isolation provided.

Example 2. In some of the complex electronic computers, redundancy is sometimes introduced to increase the MTBF.[11] One approach is to have standby units operating which are manually or automatically switched into the system to replace failed units. A simplified example of such a situation is depicted in Figure 7.3-16.

For the most elementary case, reliability enhancement is easily estimated. Let r_1 be the reliability of unit 1 or unit 1a of Figure 7.3-16, and p_1 the corresponding failure probability. If perfect failure detection and switching are assumed, the reliability of the doubly

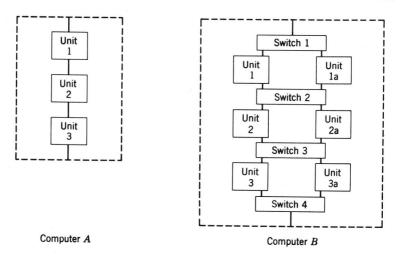

Figure 7.3-16. Comparison of series and series-parallel unit grouping of computer elements, using redundancy.

redundant system is given by

$$R_{(1+1a)} = 1 - (1 - p_1)^0(p_1)^2$$
$$= 1 - p_1^2 = 1 - (1 - r_1)^2 \qquad (7.3\text{-}10)$$
$$R_{(1+1a)} = -r_1^2 + 2r_1$$

Where failure detection and switching are imperfect, enhancement can be determined by comparing the two circuit arrangements depicted in Figure 7.3-17. This would give:

$$R_B = (R_{1+1a})r_s r_d = (2r_1 - r_1^2)r_s r_d \qquad (7.3\text{-}11)$$

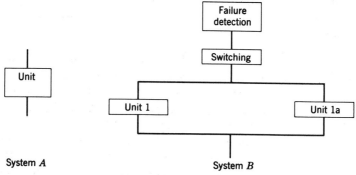

Figure 7.3-17. Alternative circuit configurations with failure detection and switching effects noted.

Designing for Reliability

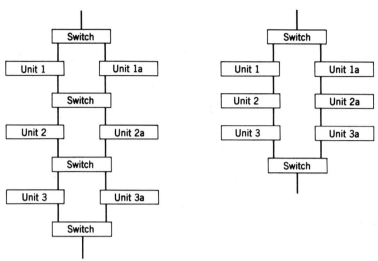

Figure 7.3-18. Comparison of alternative switching arrangements to obtain most favorable circuit redundancy.

where r_s is the switching reliability,

r_d is the failure detection reliability.

An interesting problem which arises in such cases is how to determine the arrangement of redundant parts giving optimum enhancement. For example, the two arrangements shown in Figure 7.3-18 differ very little in part complement; are they equivalent in reliability?

To answer this question, Flehinger of IBM[13] carried out an analysis for the cases of perfect and imperfect switching. Depending on the reliability of each unit as well as the degree of perfection of each switch, considerably different results can be obtained.

Strength-Stress Method. In the discussion on failure definition, reference was made to the probabilistic nature of the strength of materials and the variability of the loading. Consider then a probability density curve for the strength of a material and the associated stress probability curve from the loading of the part. A typical diagram might appear as in Figure 7.3-19. It should be noted that both the strength and the stress curves are skewed distributions in that they deviate substantially from the Gaussian ideal. The probability of failure density $f_F(x)$ is given by the product of the strength probability density and the probability of the stress exceeding that

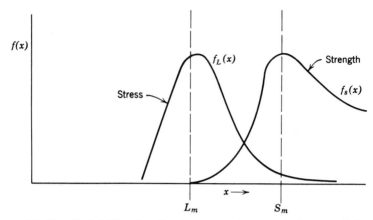

Figure 7.3-19. Probability density of stress and strength for a typical part.

strength:

$$f_F(x) = f_s(x) \left[\int_x^{+\infty} f_L(x)\, dx \right] \qquad (7.3\text{-}12)$$

Figure 7.3-20 shows a probability density of the combined distribution. The area under the curve represents the probability of failure for the specific loading and material strength:

$$P_F = \int_{-\infty}^{+\infty} f_F(x)\, dx \qquad (7.3\text{-}13)$$

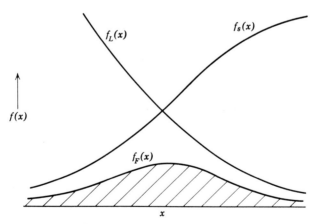

Figure 7.3-20. Probability density curve to an enlarged scale for the combined distribution of Figure 7.3-19.

Designing for Reliability 315

The important sections of the distribution curves are the normally ignored tails, where small errors can mean large errors in predicted failure rates. Thus, for this approach to the estimating of failure rates to be successful, extensive data must be obtained on the probabilistic nature of strength of materials in the subnominal range.

Experimental Test Method. No matter how good the analytical methods are for estimating failure rates, they are at best only approximations and must be reinforced by appropriate experimental testing. Although tests of overall equipments will be necessary when these items are available, tests made on suitable-sized modules are desirable when it appears from analytical studies that the reliability for these functional units will be adequate.

Environmental effects such as have been considered to be the ones that will be encountered, including 2σ and/or 3σ values, must be obtained by simulation, and the equipment performance under these conditions must be determined. Whereas stresses such as voltages, temperature, and mechanical loads can be applied directly, the effect of aging or deterioration with life may have to be simulated by using analytically obtained values or accelerated stress conditions.

The experimental test methods, though time consuming and expensive, are a necessary step to ensure good reliability estimates of failure rates.

Identify Failure Modes and Consequences

Although the concept of failure is repugnant, the facts of life teach us that all devices, be they mechanical, electrical, electronic, hydraulic, pneumatic, or otherwise, have a mortality, a probability of failure. This mortality is dictated by the relationships associating the generalized "strength" with the generalized "stress" of the device in addition to the probabilistic nature of the material properties, the chance aspects of quality control, and the influence of human error in the operation. In the concept of generalized stress (again a variability from nominal) one includes all the demands from the normal and occasional abnormal performance of the device, such as the operation in a variable environment of temperatures, vibration, contamination, and service. In the sense of generalized strength one considers the capability of the device to withstand a stress with a predictable probability of safe operation.

The definition of what constitutes a failure is an all-important task in a reliability program. The reliable operation of a device is the failure-free operation, and conversely unreliability is measured by the number of failures per unit of operating time. Thus, both sound designing for reliability and subsequent effective measurement of the achievement of reliability depend in a large measure on the concise definition of what constitutes a failure. For example, a small hydraulic leak in a fitting on the hydraulic system of an aircraft jet engine may be a failure in the sense of the maintenance definition of reliability but not in the sense of jet-engine power loss or airplane mission abort. However, under certain conditions of high supersonic flight a hydraulic leak could constitute a flight safety hazard because of the potential danger of fire in the engine compartment. Failures can thus be grouped according to their severity. A possible grouping is as follows:

1. Nuisance items.
2. Maintenance items—normal.
3. Maintenance items—special.
4. Overhaul item.
5. Component removal item.
6. Power loss event (10 per cent loss).
7. Mission abort item.
8. Inflight shutdown item.
9. Flight safety item.
10. Population safety item.

This listing is specifically written for flight propulsion devices, but similar, parallel items can be compiled for equipment ranging from light domestic appliances to nuclear reactors.

The importance of quantitatively defining what constitutes a failure is apparent. Thus, a 10 per cent loss in thrust or a 10 per cent increase in fuel consumption may refer to power loss and mission abort failures of a jet engine. The fact that an engine accelerates from idle speed to top speed in 10.0 seconds instead of 5.2 seconds may be only a nuisance item to be reported for further checks. Yet for a carrier-based jet plane an excessively slow engine acceleration may spell doom to the craft during an aborted landing operation which requires a rapid increase in thrust to prevent ditching in the ocean and to permit the pilot another try.

The quantitative definition becomes somewhat more difficult when one considers cracks in turbine discs, cracks in the trailing edge of turbine blades, or other evidences of degradation from such use aspects as thermal cycling and fatigue loading. A similar example in electronics would be performance degradation of vacuum tubes and tran-

Designing for Reliability

sistors which permits continued satisfactory operation but is a warning of increased proneness to failure.

In summary, then, a concise and complete quantitative definition of what constitutes failures is one of the essential aspects for designing for reliability. The definition of failure thus becomes a closely integrated part of the design and analysis process. We have recognized that failures need by no means be catastrophic to merit concerned attention. Consider the case of the 5 per cent loss in net thrust of a turbojet engine as a failure.[49] Thrust is a function of the following:

Jet nozzle exit area—A_8

Jet exhaust temperature—T_8

Mass flow of gas—W_8

Thrust coefficient of the jet nozzle—C_{fg}

$$F_n = f(A_8, T_8, W_8, C_{fg})$$

Thus the variation in thrust can be given by the influence coefficients in the form of partial derivatives:

$$\Delta F_n = \frac{\partial F_n}{\partial A_8} \Delta A_8 + \frac{\partial F_n}{\partial T_8} \Delta T_8 + \frac{\partial F_n}{\partial W_8} \Delta W_8 + \frac{\partial F_n}{\partial C_{fg}} \Delta C_{fg} \quad (7.3\text{-}14)$$

Each of the partial derivatives and the variations of the parameters can be associated with engine and engine control characteristics. A change in the compressor characteristics resulting from blade damage caused by the ingestion of rocks from the ground will result in a change in airflow, W_8, and jet nozzle area, A_8, for the turbine exit temperature controlled engine.

The speed of the rotor of the engine is closely regulated by a governor type of control. If a governor malfunction should cause the speed to deviate from the scheduled value, there will be an effect on thrust by a change in air flow, jet nozzle area, and possibly jet exhaust temperature (for the afterburning engine).

In essence, the above type of analysis represents the start of a failure consequence study. The definition of what constitutes a failure of a device in the gross performance sense sets forth the external evidences of failures and establishes their permissible boundaries. The failure consequence study considers the failure of a minute device inside a machine and by analysis reaches conclusions as to the consequence of such a failure (or degradation, or change in behavior) on the capability of the machine to perform its designated mission.

Since each part of a machine can conceivably fail in several ways and by several modes, the conclusion must be reached that a complete reliability analysis of a design must consider every possible failure mode of all structural and functional parts of the machine. The failure of the part must be defined in quantitative terms related to the performance of the part and of the machine, so that analytical procedures can be applied to arrive at a prediction of the consequence of such failure on machine performance.

Failure Mode Identification. If a detailed design review is performed, as opposed to a systems design review, the device under study is divided into its next smaller grouping of components and each of these components is again divided into its parts. If the study is of a systems nature, the divisions will stop at major components. The designers are primarily interested in the detailed study. Each part to be studied is designated by a code number to locate it in a block diagram structure and also to identify its nature. This information is required for further use.

Mention was made of the fact that similar parts have been known to fail in many different modes. Much benefit would be derived if one were able to call on a cumulative memory of a history of failure events, their causes and the typical trouble areas as well as the associated critical parameters. Unfortunately the design engineer's recall is not perfect, and he as an individual is not subject to as wide an experience as the industry of a specific product line. Furthermore, under the pressure of an ever-present deadline he may well not find time to consider some of these past events. The purpose of the following tables is then to supplement the engineer's memory and improve his ability to perform as complete a review as possible. The several illustrations given are only samples; they neither are complete in themselves nor do they represent completness in any one field. They do represent special interests related to the control of air-breathing aircraft engines.

The material in these check lists is divided into two categories:

(a) *The design problem check list* (Table 7.3-5). Many incidences of unreliability are caused by the inadvertent omission of important design considerations. Ideal lists would represent the cumulative memory of problem areas relative to certain typical devices. Such a list includes strength, manufacturing tolerances, assembly, environmental, operational, wear, and other types of problems. A check list of this kind must be dynamic. It must be updated periodically, be corrected, and grow with the experiences collected.

Table 7.3-5. Design Problem Check List

02 ADJUSTMENTS
  02 Adequate stop strength
  02 No seal with soft solder
  02 Sealing means provided
  02 Lockwire anchor nearby
  02 Effect of play in lockwire
  02 Vibration recalibration
  02 No snapback when released
  02 Excessive detent forces
  02 Sensitivity too high
  02 Identify adjacent adj.
  02 Clearly marked
  02 Overrun will not disassem.
  02 Overrun will not damage
  02 Vulnerability to disassem. damage
  02 Vulnerability to assem. damage
  02 Vulnerability to handling damage
  02 Vulnerability to tampering
  02 Consequences of misadjustment
  02 Tempting hand holds
  02 Accessibility
  02 Dynamic characteristic adj.
  02 Static characteristic adj.
  02 Thread fouling by set screws
  02 Min. special tools
  02 Range for service wear corr.
  02 Detent effective with eng. running
  02 Click detent for service adj.
  02 Interaction between adj.
  02 Setting changed by locking
  02 Linearity of adj.
  02 Adjustable w/o disassem.
  02 Subassemb. separately adjustable
  02 Extreme adj. indicates mfg. problem
  02 Interchangeability w/o adj.
  02 Hysteresis
  02 Repeatable contact areas
  02 Lost motion
  02 Temperature compensation

05 BEARINGS, BALL, ROLLER, NEEDLE
  05 Coking restrains lubrication
  05 Shock loading
  05 Structurally sound retainers
  05 Axial load level
  05 Self-aligning required
  05 Race retaining means
  05 Lubrication requirement
  05 Contamination protection
  05 Corrosion protection
  05 Race distortion in assem.
  05 Assem. + disassem. damage
  05 Min. special tools
  05 Assembled clearance tolerance
  05 Clearance effects on system
  05 Deflection under axial load

07 BEARINGS, FRICTION (EXCEPT CARBON)
  07 Coking retards lubrication
  07 Fracture of pressed-in bushings
  07 Shock loading
  07 Similar materials tend to gall
  07 Lubrication requirements
  07 Sleeve retaining means
  07 Shaft alignment
  07 Axial load level
  07 Clearance tolerances
  07 Effect of clearance on system

10 BELLOWS, DIAPHRAGMS
  10 Consider internal stops
  10 Internal spring rubs bellows with vibration
  10 Bellows rubs cavity with vibration
  10 Consider steel
  10 Test in operation for surge magnitude + frequency
  10 Faulty transient actions of adjacent valves
  10 Limiting leakage rates
  10 Transient pressure peaks
  10 Deflection stops
  10 Min. thickness
  10 Quality control, general
  10 Material composition
  10 Avoid soft solder
  10 No stops if liquid filled

(b) *The failure event and cause check list* (Table 7.3-6). This partial list of component failure modes, failure causes, and influencing parameters is meant to aid the engineer in preparing his reliability design review by bringing to mind some of the modes of failures experienced in the past. The following code is used in Table 7.3-6:

Table 7.3-6. Failure Event and Cause Check List

Actuator: Provide a specific amount of mechanical power.

Ball Screw Type

A	B	C
Excessive friction	Misalignment	Temperature limit
Structural failure	Scored or worn races	Vibration
	Bearing failure	Contamination
	Galling	Material selection
	Eccentric load	Part tolerance
	Differential expansion	
	Mounting shifts	
	Overload	

Hydraulic and Pneumatic

A	B	C
Leaks	Seal leakage	Temperature
Excessive friction	Contaminated fluid	Vibration
Structural failure	High transient pressure	Type of lubricant
	Misalignment	Filtration
	Worn piston	Part tolerance
	Mounting shifts	Concentricity
	Overload	

Alternator, Control: Provide power and signals as a function of input speed to other control components.

A	B	C
Drive failure	Heavy overloads	Temperature
Excessive backlash	Frozen/worn bearings	Vibration
Loss of magnetism	Temperature cycling	Drive alignment
Transient voltage peaks	Structural failure	Humidity
Shorted windings	Drive misalignment	Contamination
Open windings	Insulation deterioration, electrical/mechanical	Lubricant
Voltage, variation		
Poor regulation		
Low resistance to ground		

Designing for Reliability 321

Table 7.3-6 (*Continued*)

Motors: Supply a specific amount of mechanical power.

Electric

A	B	C
Sheared drive	Contamination	Temperature
Shorts to ground	Frozen/worn bearings	Vibration
High friction	Rubbing (rotor, field)	
Interference	Open/shorted windings	
	Single or multiple wire shorts	
	Structural failure	

Torque

A	B	C
Low output	Loss of magnetism	Handling
Interference	Magnetize chips restricting armature movement	Temperature cycling
Null point shift		Susceptability to contamination
Frozen in one position		
	Frozen bearings	Vibration
	Output linkage bind	Lubrication
	Open/shorted coils	
	Beam deflection	
	Contamination	

Pumps: Provide a specific range of flow and/or pressure as a function of input power.

Air Turbine

A	B	C
Seal leakage	Seals—wear out, cocking, rupture	Bearing OD & ID match with housing & shaft
Overspeed	Cavitation	
Vibration	Wear ring failure on impeller	Bearing preload
Sparking		Dynamic balance
Performance loss	Bearing and shaft failure	Seal design
Bucket fracture	Rubbing shrouds	Seal angular deflection
	Lubrication failure	Impeller overhand
	Loss of hydraulic load	Resonant frequency
		Seal heat generation

A. Failure event: physical phenomenon which results in loss of function.

B. Cause: typical reasons for failure event.

C. Typical trouble areas and critical design parameters.

Ideally both types of lists are kept up to date by continuous flow of data from the field, the factory, and the development test. These lists would thus truly represent the cumulative experience of a department or even an industry. In several modern engineering departments such lists are kept and processed on magnetic tapes of data processing machines. If the parts are identified by codes, the stored problem, mode, and cause listings of desired parts can be collected and printed out in a form readily usable for the design review.

In the failure mode identification phase therefore, a specific piece of a component is asked to fail in all the possible modes. These modes are obtained from lists as shown in Tables 7.3-5 and 7.3-6, from plain common sense, or from other information available. In Table 7.3-7, a sample analysis is shown with the failure mode column highlighted. The need for the inclusion of all possible problem areas in this phase of the study is obvious. The failure mode selection phase sets the framework for all subsequent efforts in this reliability study.

Failure Consequence Study. One of the technically and analytically most challenging tasks in the process of designing for reliability is the analysis of the consequences of a particular mode of failure. These consequences are expressed as the effect on the assembly (such as a valve assembly), on the component (such as the hydromechanical main fuel control), on the system (such as the temperature-control system of the jet engine) and on the engine itself, which could be expressed in terms of steady-state thrust loss at specific power settings, increases in fuel consumption, or, for the transient cases, inability to hold rotor speed during disturbance or a turbine inlet overtemperature during certain transients which would result in a reduced turbine reliability. A simplified sample of a typical failure consequence analysis is shown in Table 7.3-7.

The majority of failure modes can be handled in a straightforward manner requiring logical and only qualitative reasoning. Certain cases will demand a level of analytical proficiency in the study of nonlinear dynamic systems which is beyond the scope of all but a few engineering operations. Such treatments will not be discussed here but should be considered in the evaluation of analysis needs of the complex system. In this regard the degree of confidence the analyst has in the veracity of the analytical model, i.e., the equations representing the system, will decide the value of any failure consequence studies.

Table 7.3-7. *Numerical Reliability Analysis*

FUEL CONTROL

Ref. No.	Assembly/Element Modes of Failure	Effect On Assembly	Effect On Component	Design Considerations	Est. F/R	System Effect	Engine Effect	P.L. Effected Mx	Ml	95	90	N
2.8	Piston, press, regulating											
	Erode and wear	Leak of inlet to servo to metered fuel	None—self-correcting	Mat'l: AISI 440 stainless. Hardened edges. Delta P = 20 psi across edge. 188 lb force available when porting low press. 25.6 lb min. force avail. when porting high press. 17.7–188 lb force avail. (function of delta P) at null piston position. Movement: 0.050". Clearances: 0.0003–0.0008". 0.005" erosion showed negligible effect.	7.5	Slightly rich fuel flow	None					x
	Fracture	Leak servo to metered fuel	Rich fuel flow		0.5	Rich fuel flow	Liner damage		x	x	x	
	Seize open	Port to low pressure	Press. reg. valve open		2.0	Rich fuel flow	Liner damage		x	x	x	
	Seize closed	Port to high pressure	Same as above		2.0	600 pps fuel flow (lean)	No combustion	x				
	Seize null	Servo not ported	Delta P varies with metering area		0.5	Rich above delta = 40 psi	Liner damage		x	x	x	
					0.5	Lean below delta = 40 psi	No combustion	x				
2.9	Spring											
	Fracture	Incorrect null position	Lower fuel flow	Mat'l: 17-7 PH(C) steel. Free to rotate. Ends closed and square ground.	0.5	Lean fuel flow	Loss of thrust	x				
	Set	Incorrect null position	Lower fuel flow		1.3	Lean fuel flow	Loss of thrust	x				
	Dislodge	Jam head sensor	Lower fuel flow		0.1	Lean fuel flow	Loss of thrust	x				
2.10	Seat, spring, temp. comp.											
	Fracture	Incorrect null position	Lower fuel flow	Mat'l: Bimetallic 0.050 thick. Flat at 70°F. Flow path through seat to relieve delta P.	0.1	Lean fuel flow	Loss of thrust	x				
	Set	Incorrect null position	Lower fuel flow		4.0	Lean fuel flow	Loss of thrust	x				

323

Table 7.3-7 (Continued)

Ref. No.	Assembly/Element Modes of Failure	Effect On Assembly	Effect On Component	Design Considerations	Est. F/R	System Effect	Engine Effect	P.L. Effected Mx	P.L. Effected Ml	P.L. Effected 95	P.L. Effected 90	P.L. Effected N
2.11	Retainer, spring seat											
	Fracture	Dislodged spring	Lower fuel flow	Mat'l: AMS 4150 alum. anodized.	0.5	Lean fuel flow	Loss of thrust	x				
	Plug orifice	Incorrect null position	Lower fuel flow		0.2	Lean fuel flow	Loss of thrust					
2.12	O-ring packing											
	Loss of seal	Fuel leak O/B	Lower fuel flow and leak O/B	Mat'l: Goshen 1120 Buna N. Squeeze 20–35%. Delta $P = 740$ psi.	0.7	Lean fuel flow	Loss of thrust and fire hazard	x				
2.13	Screw adjustment											
	Fracture	Incorrect null position	Lower fuel flow	Mat'l: AMS 4150 alum. machined. Hard-coated all over. Adjusting torque 26.5 lb-in. Prevented from rotation by friction in detent, threads and O-ring.	0.4	Lean fuel flow	Loss of thrust	x				
	Strip thread	Incorrect null position	Lower fuel flow		0.2	Lean fuel flow	Loss of thrust	x				
	Plug orifice	Incorrect null position	Lower fuel flow		0.2	Lean fuel flow	Loss of thrust	x				
2.14	Detent wheel											
	Fracture	Incorrect null position	Lower fuel flow	Mat'l: 440 stainless, electrofilmed. Not a positive lock for adjust. screw.	0.5	Lean fuel flow	Loss of thrust	x				
	Wear	Incorrect null position	Lower fuel flow		0.2	Lean fuel flow	Loss of thrust	x				
2.15	Balls, detent (2)											
	Fracture	None	None	Mat'l: 440 stainless—standard.	0.2	None	None					x
	Wear	None	None		0.4	None	None					x

Establish Tolerances and Tests

A significant factor in the design of an equipment or system for reliability is the establishment of tolerances and tests to assure the engineer that the equipment as designed will meet the requirements or specifications of the system. Basically this assurance comes from the establishment of proper tolerance limits for performance and from successful tests of the ability of the equipment and systems to meet these limits. The factors influencing the establishment of tolerance limits are discussed in Chapter 9 of *Systems Engineering Tools*,[49] and the use of system simulation to allow suitable tests to check out parts of a system as well as the whole system has been described in some degree in Chapter 3 of the same book. The following material will serve to indicate for a particular example some of the steps for determining the effect of changes in performance caused by parameter changes and the way in which combinations of parameter change effects may be estimated. Also presented will be some ideas by R. H. Norris on the use of tests to prove the achievement of the reliability goals established for parts and for systems.

Effect of Parameter Changes on Performance

Variations from the nominal values of electrical, mechanical, thermal, or other quantities will produce changes in the performance of the equipment using these quantities. To the extent to which the performance may fall below an acceptable limit as a result of these changes, the reliability of the device or equipment is affected. Although there may be some doubt as to the magnitude of the parameter changes that will be encountered because of manufacturing tolerances or aging, it is possible in most devices for which a thorough design has been made to determine the effects on the performance of such changes in the parameters. The designer, in an effort to "center his design," should analyze these effects to help him establish initial tolerances and to specify the limit of life based on performance capability. Through a thorough knowledge of the equipment performance under all conditions of tolerances and aging the designer can be better assured that his equipment will perform reliably.

Two approaches for establishing the effects of parameter changes exist: (1) the method of small perturbations or "linearization" and (2) the method of exact analysis. As implied, the first method is an approximate one in which small increments are given to each of the parameters in turn and the effect of each variation on the performance in question is determined. In this way, a proportionality factor is established between small changes in each parameter and the resultant performance. The effect of having a number of parameters change simultaneously is a linear combination of the effects of the variations of each. To establish the probabilities involved, the probabilities assigned to the changes in the parameters are carried over to the probability for the change in performance.

The exact analysis method of studying the effect of parameter changes is essential when nonlinear circuits are involved or when the changes being considered are large compared with the nominal values of the parameters involved. In these cases it is generally necessary to perform complete analyses or tests of the parameter conditions that exist. With the aid of analog and/or digital computers, the repetitive calculations necessary to determine varations in performance due to changes in parameters may be effectively accomplished. In this way reliability factors may be arrived at which relate the probability of achieving a given performance to the probability of realizing a certain change in parameter. Because of the specialized nature of the exact analysis method, consideration here will be given principally to illustrating the linearized method.

Example: Triode Drifts due to Power Supply Variations

Drifts in a triode amplifier due to power supply variations can be expressed as a gain coefficient, G, that gives the ratio of the change in output caused by a change in each of the supply voltages. Equations for these gains are easily determined from the equivalent circuit of the electronic tube. Figure 7.3-21 shows a simple triode (a) and

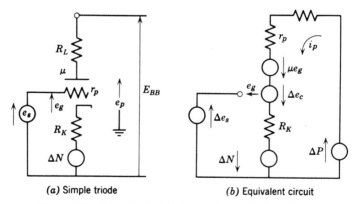

(a) Simple triode   (b) Equivalent circuit

Figure 7.3-21. Simple triode and its equivalent circuit.

Designing for Reliability

its equivalent circuit (b) in which

$$\Delta P = \text{change in positive power supply}$$
$$\Delta N = \text{change in negative power supply}$$
$$\Delta e_c = \text{change in contact potential}$$

From Figure 7.3-21

$$e_g = \Delta e_s - R_K i_p + \Delta N + \Delta e_c \qquad (7.3\text{-}15)$$

and

$$(r_p + R_K + R_L)i_p = \mu e_g + \Delta N + \Delta e_c + \Delta P \qquad (7.3\text{-}16)$$

For a triode amplifier the output voltage is measured between plate and ground. Thus

$$\Delta e_o = \Delta P - i_p R_L \qquad (7.3\text{-}17)$$

Solving these equations for the change in output voltage Δe_o, due to changes in power supplies, becomes

$$\Delta e_o = \frac{[r_p + (1+\mu)R_K]\Delta P - R_L[\mu \Delta e_s + (1+\mu)\Delta N + (1+\mu)\Delta e_c]}{r_p + (1+\mu)R_K + R_L}$$

$$(7.3\text{-}18)$$

The change in output due to change in input signal is the signal gain, G_s. From Equation 7.3-18

$$G_s = \frac{\Delta e_o}{\Delta e_s} = \frac{-\mu R_L}{r_p + (1+\mu)R_K + R_L} \qquad (7.3\text{-}19)$$

In the same manner, the gain to changes in positive supply is:

$$G_p = \frac{\Delta e_o}{\Delta P} = \frac{r_p + (1+\mu)R_K}{r_p + (1+\mu)R_K + R_L} \qquad (7.3\text{-}20)$$

Gain to changes in negative supply is:

$$G_N = \frac{\Delta e_o}{\Delta N} = -\frac{(1+\mu)R_L}{r_p + (1+\mu)R_K + R_L} \qquad (7.3\text{-}21)$$

Gain to changes in contact potential is:

$$G_c = \frac{\Delta e_o}{\Delta r_c} = -\frac{(1+\mu)R_L}{r_p + (1+\mu)R_K + R_L} \qquad (7.3\text{-}22)$$

Rewriting Equation 7.3-18 for the change in the output voltage in terms of the G's, gain coefficients, and the changes in the various voltage sources,

$$\Delta e_o = G_s \, \Delta e_s + G_p \, \Delta P + G_N \, \Delta N + G_c \, \Delta e_c \quad (7.3\text{-}23)$$

From this equation and a knowledge, real or estimated, of the Δ voltage terms on the right, one is able to determine the total change in Δe_o when changes in a number of voltage terms occur. If the maximum (3σ) or standard deviation (σ) values for these Δ voltages are known as well as the nature of their variation interrelationship—independent, random, or dependent—the combined output voltage change, Δe_o, can be estimated as illustrated below. It is apparent that changes in other parameters, such as R_L, r_p, R_K, and μ, can also affect the output voltage or its sensitivity, and hence still other gain coefficients, G's, could be developed as desired.

Combination of Parameter Change Effects

Parameter change effects tend to fall into two categories, random and systematic or bias. The *random* effects are ones in which there is no correlation or order between the changes in one parameter and those in another. The individual component changes in performance due to these parameter changes can be combined as the square root of the sum of the squares.

The *bias* effects are ones caused by known physical laws which always occur under a given set of circumstances. If two power supplies are served by a common transformer, a drop in the voltage of this transformer will cause changes in each power supply that are related to one another. The individual component changes in performance resulting from the bias parameter changes must be summed (with appropriate sign) to obtain the net bias change.

Thus, if

$$\Delta e_o = G_{r1} \, \Delta e_{r1} + G_{r2} \, \Delta e_{r2} + G_{r3} \, \Delta e_{r3} + G_{b1} \, \Delta e_{b1}$$
$$+ G_{b2} \, \Delta e_{b2} + G_{b3} \, \Delta e_{b3} \quad (7.3\text{-}24)$$

where Δe_{r1}, Δe_{r2}, and Δe_{r3} are random and Δe_{b1}, Δe_{b2}, and Δe_{b3} are bias quantities, the resultant value for Δe_o is

$$\Delta e_{oR} = \pm \sqrt{(G_{r1} \, \Delta e_{r1})^2 + (G_{r2} \, \Delta e_{r2})^2 + (G_{r3} \, \Delta e_{r3})^2}$$
$$\overline{+ [G_{b1} \, \Delta e_{b1} + G_{b2} \, \Delta e_{b2} + G_{b3} \, \Delta e_{b3}]^2} \quad (7.3\text{-}25)$$

or in a more general form

$$\Delta e_{oR} = \pm \sqrt{\sum_1^n (G_{rn} \, \Delta e_{rn})^2 + \left(\sum_1^n G_{bn} \, \Delta e_{bn}\right)^2} \quad (7.3\text{-}26)$$

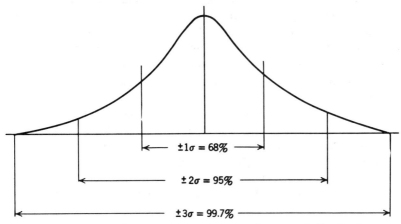

Figure 7.3-22. Approximate per cent of area under normal distribution curve for 1, 2, and 3σ.

Depending on whether the maximum values for the Δe's are used for the parameter changes, or the standard deviations are used, the value for Δe_{oR} will correspond to the maximum or standard deviation. Assuming that the overall distribution of these changes is Gaussian, as shown in Figure 7.3-22, the maximum (3σ) value for Δe_{oR} corresponds to 99.7 per cent of the cases. Two-thirds or less of this value of Δe_{oR} will be obtained in 95 per cent of the cases, and the value of change will be less than one-third of Δe_{oR} 68 per cent of the time. If the standard deviation values of the Δe's are used, the value of Δe_{oR} will be less than that calculated 68 per cent of the time.

With a knowledge of relationships such as those given by Equations 7.3-21 through 7.3-25, and of the 1σ, 2σ, and 3σ values from Figure 7.3-22 for the various equipments for which tolerances must be established, the designers are in a position to apportion to the individual parts feasible and reasonable shares of responsibility for approaching and meeting the overall performance goals sought.

Tests to Prove Achievement of Reliability Goals

Success in meeting the "reliability demonstration" requirements of a government contract is of course the object toward which much of the reliability effort is directed. The reliability test requirement for electronic equipment which has come into widespread use in that specified in U.S. Air Force Specifications MIL-R-26667A. It is based on the recommendations of "Groups 2 and 3" in the AGREE[10] report.

The nature of the AGREE recommendations for reliability tests has been summarized by Holahan[30] as follows:

"The AGREE reliability procedures are based on the pretesting of sample lots of components and end equipment under specified environments, and on mathematical interpretation of the results. Where failures occur, their cause is determined and then removed by remedial action, and the unit is tested again.

"At the outset of a program, a reliability goal is worked out for the overall system and then, by an apportioning process, for subunits and components. Component selection therefore is one of the most important phases of the AGREE procedure.

"The AGREE tests are based on statistical methods. After a certain number of samples has been tested prior to production, the test time and the number of failures (both easily measured) are inserted in probability equations along with *specified* 'mean time between failures' (t_1), *minimum acceptable* 'mean time between failures' (t_2), the risk of error, and the desired confidence level of the resulting decision about reliability of the equipment. The equations yield a *probable* 'mean time between failures.' If this figure turns out to be below the acceptable minimum, the reason for the deficiency is tracked down and the equipment is brought up to the required level (by correcting defects in workmanship, substituting better components, or redesign). During the production run, samples are taken at specified intervals to check the MTBF.

"Essentially AGREE procedure relies on sequential testing to check which of two hypotheses applies:

MTBF $\geq t_1$ (the actual MTBF equals or exceeds the *specified* MTBF); or

MTBF $\leq t_2$ (the actual MTBF is equal to or less than the *minimum* acceptable MTBF).

"Obviously, if one hypothesis applies, the other doesn't. Testing is continued until the results show beyond doubt (at the required level of confidence) which hypothesis applies. The total test time therefore does not have to be fixed. [Figure 7.3-23 presents one example.]

"To prevent prolonged stay in the 'continue testing' region, the curve is truncated [at points beyond the range of Figure 7.3-23]. The truncation lines set by AGREE are determined by $t = 31t_1$ units and 41 failures (where t = accumulated test time)."

For further details, too extensive to present here, see References 27 and 30 and the AGREE report[10] itself.

The above type of test requirement is intended to apply primarily

Designing for Reliability

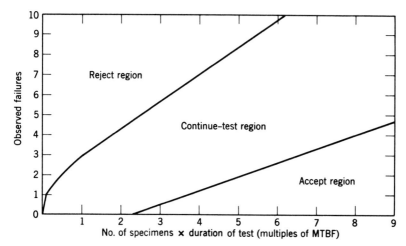

Figure 7.3-23. Curves showing reject region, continue-test region, and accept region for number of observed failures versus number of specimens times tests.

to electronic systems and perhaps also to aircraft power plants, for which an occasional failure (at, say, intervals of the order of 10–1000 hours) can be tolerated, that is, where replacement of the failed part of the module is often a practicable maintenance operation. On some equipment, however, such maintenance replacement is not physically possible. Other more stringent test methods are necessary for these cases.

Further mention of test methods to prove the achievement of reliability goals is made in the following section, where the problem of controlling quality in reliability and some methods of testing are presented.

Review Design and Feedback Results

Design reviews are an organized method of providing an opportunity to put into practice a number of old saws, such as "hindsight is better than foresight," "two heads are better than one," and "it pays to ask an expert his opinion." They are being used more extensively by reliability-conscious systems engineering groups and are specifically required by many military contracts. Design reviews entail the examination of designs for feasibility, performance, reliability, and producibility by a team of specialists and consultants representing the maximum possible experience and competence available in the are under consideration.

The design review is intended to accomplish the following:

a. Evaluate design concepts and approaches to assure optimization of the available engineering and manufacturing skills.
b. Evaluate whether chosen designs meet reliability and maintainability objectives.
c. Establish the inherent reliability of the design by review of predictions, failure modes, and failure consequences, and evaluation of engineering test results.
d. Identify potential problem areas and make the necessary recommendations for development and reliability programs.
e. Review and evaluate equipment performance and producibility.
f. Furnish management with tools to assist in making decisions on critical technical problems.
g. Provide a quantitative measure of risk regarding contractual performance, reliability, schedule, and maintainability commitments.

Scope. The utilization of design reviews is a highly desirable procedure for all new design programs and redesign and improvement programs which require requalification of the unit or system. Requirements for design review should be clearly shown and scheduled on all program PERT charts, and the time and funding necessary for their implementation should be provided within the program plan. The number and types of design reviews may vary depending upon program length and scope, but in general there should be no less than two reviews of the types categorized below.

Although it is anticipated that the number of reviews will vary from one task to another, the reviews held should be conducted at specific monitoring points of the program, such as described in the following outline.

PROPOSAL REVIEW. This review will evaluate the proposed and alternative design approaches suggested to meet the customer objectives, the relative technical feasibility, the producibility and risks involved, the trade-offs in performance, cost, and reliability, and a review of exceptions to the customer's requirements and specifications.

CONCEPT OR SYSTEM PLANNING REVIEW. This review will entail an examination of the design criteria, the environmental and operational requirements in the areas of performance, reliability, and maintainability. It will examine and discuss the technologies, feasibility, trade-offs, and possible difficulties in achieving the required objectives. It is anticipated that, as a result of this review, supporting investigations and tests may be recommended, and special problem areas identified for closer monitoring throughout the program.

Designing for Reliability 333

PRELIMINARY LAYOUT OR ELECTRICAL FUNCTIONAL DESIGN REVIEW. It is anticipated that, at this review, a preliminary reliability analysis will be available, specific test programs and approaches will be defined, trade-offs in the areas of reliability, value, and performance will be presented, and manufacturing feasibility should be established. A failure effects analysis should have already been made, and a summary of potential critical part failures should be available for this review.

FINAL LAYOUT OR EXPERIMENTAL BREADBOARD MODEL REVIEW. At this stage, the design review should be primarily concerned with (1) outstanding difficulties uncovered as a result of breadboard operation, (2) the adequacy of vendor reliability programs, (3) the adequacy of assurance and qualification test programs, and (4) a reiteration of system and equipment objectives and anticipated results.

DETAIL DESIGN OR PROTOTYPE DESIGN REVIEW. This review should be strongly oriented toward (1) minimizing subsequent production difficulties, and (2) assuring complete adequacy of the design to meet customer requirements as demonstrated by review of evaluation program, analysis of failures, qualification demonstrations, adequacy of second sources (a manufacturing responsibility), and (3) recommendations for process and design changes deemed necessary before final production.

PRODUCTION RELEASE REVIEW. This review will cover any last-minute changes, the adequacy of corrective actions taken as a result of prior reviews, a review of all open problem areas from prior design reviews, the adequacy of the failure analysis programs, standing instructions for quality control, process instructions, and the adequacy of continuing reliability controls.

Appointment of Design Review Board. Although there are many different ways in which design review boards may be organized, the following method has been found to be effective for a large engineering department doing military work.

A permanent design review board will exist for each major product line. The board will consist of five permanent members, plus the systems project engineer and as many additional temporary members as will be required for expert appraisal of the disciplines and factors covered at each of the planned reviews. The five permanent members will be:

The design review systems engineering representative;
The design review reliability engineering representative;

The design review manufacturing representative;
The design review quality representative; and
The design review product service representative.

The method of selection of the permanent members and their responsibilities are outlined as follows.

1. DESIGN REVIEW SYSTEMS ENGINEERING REPRESENTATIVE. This will be the designated reliability engineer or reliability specialist in the systems engineering unit. The systems engineering representative is charged with the responsibility for the planning and scheduling of design reviews. In this capacity, he is responsible for establishing, with the concurrence of the design review reliability engineering representative and the systems project engineer, the selection of the temporary members of the review board, the requirements for documentation, and the source material required at each review phase, e.g., work statement, design criteria, environmental criteria, block diagrams, reliability predictions and allocation of components selection criteria, worst case analysis, etc. It is the responsibility of the systems engineering representative to assure that documentation is available three working days before the review and that distribution is made to all review board members. The systems engineering representative will, with the assistance and concurrence of the reliability engineering representative and the systems project engineer, prepare an agenda for each design review meeting.

2. DESIGN REVIEW RELIABILITY ENGINEERING REPRESENTATIVE. This will be the designated reliability engineer in the design reliability and maintainability unit having functional design reliability responsibility for the product line under review. The reliability engineering representative has a dual function as a member of the board. He is (1) a technical consultant in reliability techniques and, therefore, responsible for making available to the systems engineering representative the appropriate documentation in the areas of reliability requirements, prediction, failure effects, etc.; and he is also charged with (2) a controlling and auditing function. In this latter capacity, he will—

   a. Upon completion of a review, issue a formal report to all review participants and to the appropriate product line management delineating:

      (1) Areas of uncertainty or high risk established as a result of the review and those individuals responsible for initiating further investigation or action in these areas;

(2) Recommendations of the review board; and
(3) Action plans agreed upon and the individuals responsible for their implementation.

b. The reliability engineering representative, as part of the controlling function, will also issue a monthly status report on items $a(1)$, $a(2)$, and $a(3)$, indicating for each of the above actions taken whether the item remains open, or whether it has been satisfactorily closed. All open items will be reported on a monthly basis, and all closed items may be dropped once the reason for closing has been reported in the status report.

3. DESIGN REVIEW MANUFACTURING REPRESENTATIVE. The product line manufacturing manager will appoint a permanent representative to the review board whose function it will be to provide special competence in the areas of manufacturing, including schedules, producibility, and second sourcing.

4. DESIGN REVIEW QUALITY REPRESENTATIVE. The manager of quality control will appoint a permanent representative to each product line review board to provide special competence in all areas of quality control.

5. DESIGN REVIEW PRODUCT SERVICE REPRESENTATIVE. The manager of product service will appoint a permanent representative to each product line design review board to provide special competence in all areas of product service.

In addition to the five permanent members of the design review board it may prove effective to have an additional one or two persons with special competence either through training, experience, or interest who are well qualified to question, advise, or recommend methods for improving the design and its associated procedures. Depending on the particular design being reviewed, the choice of specialists or consultants may include either in-house or outside individuals who can provide the specific type of information appropriate for the equipment or system being reviewed.

Review Reports. Minutes and recommendations of each design review meeting are expected to be issued within three working days after completion of the review. It is the responsibility of each of the designated permanent representatives in the areas of systems engineering, manufacturing, quality control, and product service to provide information on the status of recommendations and action plans in their specific functional area to the reliability engineeirng representative for inclusion in a monthly status report.

Authority. The nature of the design review board is advisory to the functionally responsible product team and is not intended to provide a substitute for the thorough creative competence of the engineer charged with design responsibility. However, the cognizant design engineer should give appropriate consideration to the recommendations of the Board and utilize the experience and competence of its members to the fullest extent possible. Reasons for nonconcurrence with the board's recommendations should be documented and forwarded to the reliability engineering representative for inclusion in the status report.

In addition to the use of the design review board during the proposal, design, and production phases of an equipment or system, it should be brought together periodically to review the record of experiences of the system as judged from results fed back from actual field operating conditions. Through such use of field reports, it may be possible to detect areas of potential later difficulty either for this design or for later designs which might otherwise be patterned on the same principles as were used in the present design. Recommendations for such changes as appear appropriate should be made for consideration by the responsible design engineer.

7.4 Controlling Quality for Reliability

Referring to the basic definition of reliability, one notes that "Reliability is the probability that a part or system will operate for the use intended in the environment encountered." This probability can be considered as being composed of three parts: P_I, the inherent reliability of any one of this kind of part if it were built to the design specifications; P_{MA}, the reliability of each particular part as manufactured and assembled, starting with these design specifications; and P_{OM}, the reliability of a part as designed and manufactured, and as operated and maintained under the environment actually encountered.

Figure 7.4-1 illustrates pictorially the significance of each of the three reliability interpretations noted above. As can be seen, to the extent to which the combined $P_I \cdot P_{MA}$ product is less than the inherent design reliability, P_I, alone, there is a need for improved control of quality in the manufacturing process.

In Section 7.3, the emphasis has been placed on getting the inherent design reliability to be adequate for the purpose intended. In the material of this section, influenced strongly by the work of R. H. Norris, attention will be given to some aspects of controlling quality

Controlling Quality for Reliability

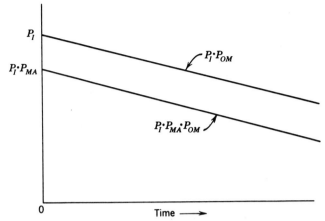

Figure 7.4-1. Representation of the reliability terms P_I, P_{MA}, and P_{OM} as a function of time.

to enhance P_{MA}, such as the following:

Production volume in large numbers uninterruptedly.
Cleanliness and process control.
Information collection on failure incidents.
Investigation to diagnose the cause of the symptoms.
Invitation and execution of corrective action of normal failure occurrence.
Use of screening methods and "burn-in" parts.
Coordination between components of the organization.

Production Volume in Large Numbers Uninterruptedly

The object of large production volume for reliability applies to both the parts and materials which are purchased and the products which are made in-house.

First the matter of production in large numbers will be considered. When the required failure risk is very low (e.g., below 0.1 per cent per 1000 hours, as is increasingly the case with the present trend to complex systems), this reliability can be demonstrated only when the production process continues over a relatively long time, in large numbers. Only by tests on large numbers of specimens can *evidence* be provided of whether or not the requirement is achieved. For example, consider a test lasting 1000 hours, and a lot of 1000 specimens. A failure of just 1 specimen would indicate a failure rate of 0.1 per cent per 1000 hours (at the conditions chosen for the test), but with a very low confidence level (i.e., very low statistical significance). For a lot of 10,000 specimens, which yielded 10 failures, the same failure rate would be indicated, but with much more statistical significance. When failure rates approaching 0.001 per cent per 1000 hours are required, the number of specimens needed to provide significant evidence in a 1000-hour test approaches a million! Such a number is of course impractical.

Furthermore, steady production, without interruptions, is needed, for evidence of low failure rate will not be found unless practically all sources of flaws, contamination, and abnormal processing have been eliminated in the production process. If a process is stopped and then restarted after a considerable time interval, there is opportunity for abnormalities in processing due to changes in the adjustments of the processing equipment, or abnormalities in chemical composition, to occur, and the effects of these abnormalities may not be detectable until large numbers of specimens have been tested for a long period of time.

Next, consider the production of large systems, such as jet engines. Here the number of possible sources or "modes" of trouble is far greater than in a small part such as a transistor. Elimination of one source of trouble on one engine does not prevent a different trouble from occurring on a second engine. In fact, it has been suggested that the state of progress in "debugging" a new design of engine be judged by a plot of the frequency of occurrence of *new* modes of failure (i.e., modes not previously encountered), as successive engines are produced and the number of operating hours rises on each. Only after production has proceeded for a relatively long time, can the chance of new (and hence not yet remedied) modes of failure become tolerably low.

Cleanliness and Process Control

Dirt or contamination in other forms is one of the commonest sources of failure of parts of complex systems. Sometimes dirt enters in the processing of the materials, sometimes in the assembly of the parts and of the system, and sometimes during the test procedures. Furthermore, the hazard is very often erratic in its occurrence, so that early reliability tests may fail to detect it.

How can dirt be effectively kept out? The answer depends on the degree of cleanliness required as well as on the source of the dirt. Every product has its own cleanliness requirements. However, these requirements can be classified into several groups, as follows:

Class A or *"ultraclean"* room requirement, as has been found necessary for surgical operations, manufacture of transistors, miniature bearings, and biological products. The requirement of gauze masks for mouth and nose (as used by surgeons), at least for persons suffering from head colds or hay fever has also been suggested.

Class B or *"superclean"* area requirement, as has been used in manufacture of certain electronic and optical instruments.

Class C or *"clean"* area requirement, as has been used for assembly of watches, cameras, and refrigerator compressors.

Sometimes the chief hazard from dirt is not in the general environment, but in the liquid or gas that the product utilizes (e.g., fuel or lubricant). In such cases, simple screens or more elaborate decontaminating processes may be required to reduce failure risk from such sources of contamination. Common examples of such hazards are closely fitting valves and pistons in hydraulic control systems and in refrigerator compressors. Fuel for jet engines is notoriously subject to various unpredictable impurities which tend to cause "gumming" of tubing surfaces at elevated temperatures.

Sometimes dirt enters not during the manufacturing process itself, but during subsequent inspection or test procedures. For example, during a certain test procedure a mechanical part made of magnetic steel attracted tiny particles of iron dust which were not carried away by the subsequent mild dust-removal operation. Another example is chemical contamination of one part from the fumes of solvents used for cleaning purposes on an adjacent part.

Process control is a large and important problem, primarily of concern to manufacturing rather than engineering personnel. Although no attempt is made here to discuss this subject, it is worthy of serious attention so that consistent processes known to be satisfactory can be assured.

Information Collection on Failure Incidents

When a new product is being developed, a "debugging" process is customarily necessary. This process may last many months, since some troubles do not become evident immediately or on the first prototype specimens. Some malfunctions develop only during use by the customer. Government contracts sometimes require periodic reports of all such troubles.

Information which should be collected includes the following:

a. The "symptom" of the trouble should be identified. When the trouble occurs during use by the customer, it is unwise to seek a "diagnosis" of the cause from the customer's technician. He may not be competent to make a diagnosis! However, he is generally able to describe the symptom, namely, the nature of the malfunction which he observes. He should be asked to do this in his routine report of the failure incident.

b. The part, module, or component responsible for the malfunction should be identified and replaced. Often it is more practicable to replace the module than to replace the part. If the replacement is successful in remedying the malfunction, the item replaced not only should be identified in the report of the malfunction but also, when practicable, should be returned to the manufacturer for diagnosis.

c. The duration of accumulated operating time previous to the trouble should be reported also, to provide evidence on the question of whether the failure is in the "infant-mortality" region, the "random-failure" region, or the "wear-out" region. The remedial action to reduce the risk of such failures in the future varies, depending on which region is encountered. To provide a basis for collecting the information here desired, it may be necessary to provide "elapsed-operating-time" recording devices.

Investigation to Diagnose the Cause of the Symptoms

The hardware returned as defective can be exploited to improve future reliability in the following two ways:

1. Tests to *identify which part* is responsible for the malfunction, and then inclusion of this fact in the statistical compilation of failure rates of that type of part.

2. Investigation to *diagnose the cause* of the failure of that part, e.g., a mechanical defect or presence of an abnormal impurity. This may well be followed by further investigation to identify the cause for that defect or impurity.

The first of these methods is relatively simple. It permits—when sufficient evidence has become accumulated—a revision (or confirmation) of the failure rates of the various types of part. Large revisions of initial choices of failure rates may sometimes be initiated by such evidence. These rates, when realistically evaluated, are useful to the design engineer to guide his choice of alternative types of parts, alternative circuits, alternative provisions for heat dissipation (to reduce

Controlling Quality for Reliability 341

failure rates by decrease of part temperature), and other means for meeting the reliability requirements.

The second of the above two methods is more difficult to accomplish. However, it is likely to yield greater improvement in reliability, since it can be used to identify the point where corrective action is needed—in fact, it is generally an essential prerequisite to such action.

Initiation and Execution of Corrective Action

Corrective action for failures of a "normal" occurrence can better be understood through the classification of defects or failure sources of parts into the following two types:

1. Defects of "normal" distribution, i.e., those which are a common hazard in the "normal" population of such parts. Any characteristic property of a part, such as rate of drift of resistance of a resistor with time, or the strength of a structural part, is likely to show appreciable variation from specimen to specimen. Such variations often are found to be representable by a "normal probability distribution" about some mean value. If the frequency of occurrence of a defect is reasonably consistent with such a distribution, it is here called a "defect of normal distribution."

2. Defects which are "abnormalities," *not* present in the "normal" population of the parts. If the defect is of a nature *not* to be expected from the normal distribution of the variations from specimen to specimen, it is here classified as a "freak" defect. For example, a factory worker may sneeze just once and spray tiny droplets of "sneeze acid" on a particular part. This impurity may in time cause enough corrosion to result in malfunction of the part. But the impurity would be completely absent in all other specimens of that part. Often such freaks occur so rarely that the origin cannot be identified.

Corrective action such as closer tolerance limits or a change of material or of dimension is therefore generally not practicable for this type of failure source, since other freak defects may occur in spite of the change. The only remedy for freaks seems to be the use of screening methods to catch them before they are incorporated in the product. Such methods are discussed below.

The corrective action here mentioned is therefore the action directed to prevention or reduction of only those defects which are found in the "normal population." No attempt is made here to consider in particular how such action is carried out; it is a normal procedure in the development of any new product.

Use of Screening Methods—to Remove Abnormal Parts and Materials

The method of screening which gives high likelihood of catching any freak defects that may exist in a given "lot" of parts for electronic systems, even when it is not known what the nature of the defects may be, is called "rugged run-in."

What is "run-in," or "burn-in" as it is also known? It is the operation of a part, at some chosen temperature and stress conditions for some chosen period of time, before assembly of the part in the final product. The purpose is to discover and screen out those specimens (of a purchased lot) which are weak enough to fail during this run-in. A growing body of evidence shows that the failure rate of most electronic parts decreases with increasing time for a long period, e.g., up to 20,000 hours at moderate temperatures. Hence, the longer one runs these parts, the lower the failure rate will be thereafter, provided one does not encounter an appreciable risk of failure from "wear-out."

Proposed Method for Selection of Stress and Duration for Run-in. The first step, to permit proper use of the run-in method, is to make tests to establish the wear-out limits of the part as functions of applied "stress" and *time* (or, for fatigue-stress and for erosion of relay contacts, the *number* of cycles). Figure 7.4-2 from R. H. Norris illustrates typical results of such tests for a silicon transistor.

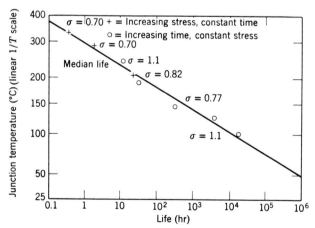

Figure 7.4-2. Arrhenius-type plot of test results for median life.

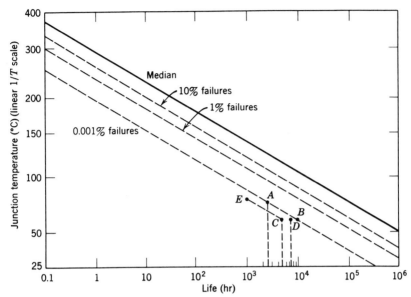

Figure 7.4-3. Arrhenius-type plot, showing life for various percentiles of the population.

The "stress" in this case, is the reciprocal, $1/T$, of absolute temperature, T. This particular type of stress was chosen for this correlation for the following reason. It has been well established that, when "wear-out," or degradation, of a part is caused by slow chemical reaction or by slow diffusion (of one material into an immediately adjacent material), this "stress," $1/T$, can be represented approximately by a *linear* function of the logarithm of the time, τ, required for a given amount of the degradation to occur, i.e.,

$$\frac{1}{R} = c_1 + c_2 \log \tau$$

where c_1 and c_2 are empirical constants. This equation is one form of the so-called "Arrhenius' relation."

This equation is represented, of course, by a straight line when plotted as in Figure 7.4-2. But the variation from specimen to specimen of the lot under test produces a probability distribution of the test results about the "median" of the lot. Use of the test results permits drawing the lines for various percentiles, e.g., 10, 1, 0.1 per cent, etc., as shown in Figure 7.4-3.

The second step is to designate the desired service life of the part, such as 2500 hours, and also the fraction, such as 0.001 per cent, which can be allowed to fail by such degradation by the end of the chosen period of service. The result is shown, for the examples specified, by point A in Figure 7.4-3.

The third step is to select a factor by which to multiply the desired service life, to allow for a considerable fraction of the allowable degradation to be "consumed" during a run-in process on the part before its release for use or reliability demonstration. Let us select, for example, a factor of 4, thus obtaining a time of $4 \times 2500 = 10{,}000$ hours. We can then allocate, say, 5000 hours for run-in, 2500 hours for actual service, and the remaining 2500 hours as a margin for inaccuracy in the derivation of the degradation curve of Figure 7.4-3. For this time of 10,000 hours, the point of the line in Figure 7.4-3 is here designated as B and the corresponding temperature as T_2. Hence T_2 is the allowable maximum temperature of the part in service. Cooling must be provided sufficient to keep the part temperature below T_2. But this T_2 is not impracticable; it is about 60°C.

The fourth step is to select a higher temperature for the run-in screening process than this T_2, for perhaps we cannot wait 5000 hours for the completion of the run-in. Suppose that we can afford, for example, only 1000 hours. The slope of the curves in Figure 7.4-3 shows that the amount of degradation will remain the same if the temperature is 79°C for 1000 hours as it will at 60°C for 5000 hours.

The result is, for this example, that we run-in the parts for 1000 hours at 79°C (to point E in Figure 7.4-3). Then those parts which survive are put into service for the intended 2500 hours at an allowable maximum part (junction) temperature of only 60°C. This procedure ensures that we have not produced too much "wear-out" by our choice of a rugged run-in process.

Need to Include Temperature Cycling in the Run-in Process. There are other causes of failure to consider also. One is temperature cycling, which may result in thermal fatigue failures of the materials, caused by unequal coefficients of thermal expansion. The run-in process can be modified to test susceptability to this hazard also. Let the run-in process be not steady, but in a series of successive alternations of warm (79°C) and cool conditions. The cool temperature should be below the warm temperature by enough, call it ΔT, to slightly exceed the maximum fluctuation to be expected in service. The number of cycles should also slightly exceed the maximum number to be expected in service. Large margins of safety for this ΔT and for the number of cycles are probably undesirable, since they might lead to excessive thermal fatigue damage, not representative of the worst service-life conditions. But if the run-in tests should, by chance, show an appreciable percentage of the parts to fail by thermal fatigue, this will indicate serious danger of further trouble from this source in service. The allowable temperature range in service should then be reduced, and the run-in tests repeated, or else a different type of part should be tried.

Other screening procedures are available but will not be described here.

Coordination between Components of the Organization

The difficulty of achieving satisfactory coordination between various organizational components of a business is often one of the major hindrances to achieving progress in reliability.

The chief obstacle seems to be an unrealistic, but widely prevalent, attitude of management. This attitude is that an engineer should be able to specify by words and drawings just what he wants, from use of written information furnished him, without personally inspecting pertinent hardware or participating in tests. In a rapidly advancing industry, such as modern products for the armed services, this is just not possible.

In some industries, aircraft for example, this obstacle is overcome by providing an engineering team to carry a major project all the way from the proposal stage through system specification, design, prototype manufacture, acceptance testing, and first manufacture. The team then goes back to proposal work to start the next round. This often works reasonably well.

In other industries, however, there is no single man, let alone an engineering team, that follows a project through from start to finish. The engineers who plan the system, and the designers, may never see the hardware resulting from their efforts, and they may have no effective feedback on how it behaves under test. Words and numbers on paper are not sufficient; they must see the hardware and observe its performance with their own eyes. This is particularly true when, as is frequently the case, the hardware proves to be not as satisfactory as desired. It is even more true when the trouble occurs only after many hours of operation, that is, where reliability is involved.

To achieve reliability it is *essential* that all engineers participating in a given project be actively aware of the effect their individual actions have on the ultimate reliability of the product. This requires close coordination and communication throughout the project, so that designers see hardware, participate in tests, and get feedback on service behavior. Only in this way can design principles, features, and techniques be developed that have proven, reasonable likelihood of producing satisfactory reliability in the product.

The coordination between different components within an engineering section depends on how the section is organized. One common practice is to provide a "project engineer," who is responsible for the coordination on one particular project. To be effective, this approach requires continuity, of course; the same individual should retain the position of project engineer from the initial concept of the project through to the final manufacture. This approach also requires for effectiveness that the project engineer frequently inspect the actual hardware and the test equipment, and consider their adequacy in the light of his own technical judgment. He cannot rely solely on the words of others.

A different, rather novel, approach to the coordination problem is to appoint for any one project a team of three men, one from engineering, one from manufacturing, and one from marketing. They report, as a group, to the general manager for the duration of their project. Their responsibility is for the overall running of the project, including its reliability aspects. They serve the same function as a project engineer but can provide broader coverage.

Quality control of the manufacture and assembly of parts of a system as well as of the system itself is a decisive factor in achieving reliability. Although not specifically the responsibility of the systems engineer, he must make it his business to see that the necessary steps to ensure the specified quality are taken and maintained in the factory, using principally the methods of good judgment and the art of "friendly persuasion."

7.5 Maintenance for Reliability

Maintenance is another factor which can have a very significant influence on the success of a system and its reliability but for which engineering does not have an active role in enforcing performance. However, engineering does have the responsibility for including provision for maintenance in its overall concept for the utilization of the system and for instituting the necessary steps by which a well-organized maintenance plan is incorporated into the broad plan for system implementation.[42] The organizational aspects of systems engineering in Section 2.1 have served to show the means whereby the effects of maintenance are included in engineering planning.

The influence of a maintenance policy on engineering design, and vice versa, can be brought out by considering two basically different approaches to high-reliability operation, namely, *no-repair* versus *preventive maintenance*.

The term no-repair operation describes the traditional high-inherent-reliability approach to operation in which the basic equipment design has a sufficient margin of safety built-in so that it can function over the entire premission operation and full mission operating periods. For example, if 100 hours of preoperational use and 20 hours of mission operational use are required, the equipment might be designed for an operational life of 600 hours or more for 90 per cent of the equipments tested. Figure 7.5-1 illustrates the stress-time relationship for such a condition in which the stress capability of the equipment far exceeds the operating stress requirements.

The term preventive maintenance operation describes the reliability approach employed in maintaining aircraft engines, where engine availability is provided by a periodic maintenance and overhaul proce-

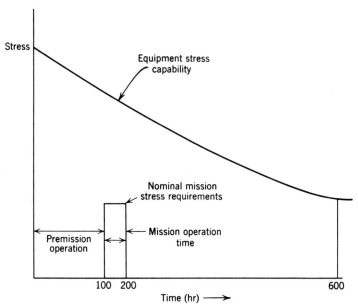

Figure 7.5-1. Stress capability versus time for equipment and mission with a no-repair maintenance policy.

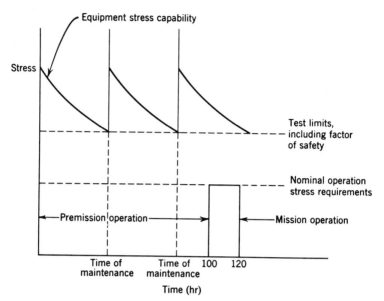

Figure 7.5-2. Stress capability versus time for equipment and mission with a preventive maintenance policy.

dure employed to guard against equipment deterioration with operation. For example, with 20 hours mission operational use and 100 hours preoperational use needed, the equipment might be such that it cannot be better designed than for a reliable operational life of 60 hours or more for 90 per cent of the equipments tested. The operating instructions must then provide for periodic inspection and replacement at intervals of, say, 45 hours of use. Further preventive maintenance must be done at such times that it is possible to ensure that during the 20-hour mission operation a 45-hour operating period will not be exceeded, with some margin of safety. Figure 7.5-2 illustrates in a general fashion the stress-time relationships involved in this type of preventive maintenance approach.

Once having established whether a no-repair or a preventive maintenance approach should be employed for the various systems, it is important that a consistent set of procedures and limits be established for successive tests. Starting with subsystem tests, the testing severity or accuracy requirements are designed to be in excess of the nominal operating requirements of the mission and to be gradually less stringent as the check-out proceeds toward final tests. Figure 7.5-3 illustrates how the progressive reduction in testing severity might be scheduled through succeeding check-out steps. In order to accom-

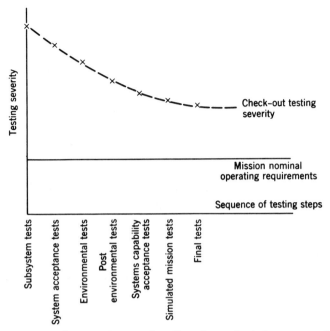

Figure 7.5-3. Testing severity as a function of a graduated sequence of testing steps.

plish such a graded program as this, an overall plan for relating the testing severity in the various locations involved is necessary.

From the preceding it is apparent that the detailed features and test values of the check-out plan will be dependent on the nature of the physical characteristics of the vehicle equipment and systems involved. It is important that the equipment design and the test and maintenance design be considered concurrently so as to enable the work on both to be most meaningful.

A consistent and thorough maintenance procedure which is rigidly adhered to or is changed only after a comprehensive study is essential for good reliability results to be obtained. It is engineering's responsibility to provide a design for such a suitable maintenance program and to outline the steps whereby it can be implemented by those responsible for this function.

7.6 Management with Reliability in Mind

Reliability is the result of an overall attitude on the part of the whole organization responsible for the system, from its general manager through all the individuals associated with the project down to the lowest-level job. Dirt carelessly "swept under the bench" by a person assigned the job of cleaning the floor may ultimately find its way into a place where it can impair the operation of a mechanism vital to the overall system. Reliability is not something that can be accomplished once and will last forever. It must be paid for continually with the price of eternal vigilance. Management must demonstrate its leadership and willingness to support its people who are trying to do the kind of job that high reliability requires.

Three Major Items

Emphasis on higher reliability requirements in recent years has required that the *top management* of the organization be involved in the support of the reliability goals of the system, objective, or project. To this end, top management must be provided with information regarding the reliability problems and their practical significance on each proposed contract, including such things as:

1. The *quantitative* nature of the reliability requirements in the specifications of the contract (as measured, for example, by the failure rate for the average part).
2. The extent to which these are more severe than the best achievements of the department to date.
3. The opinion of experienced reliability personnel, and others, on the feasibility of achieving these new requirements, and with what expenditures of time, personnel, and money. Such information must be provided in order to obtain the support of top management if additional specialized personnel, with the attendant increase in expense, are needed to achieve the required reliability.

Top management must also be aware of the long-term consequences of failure to achieve the reliability requirements imposed by modern systems specifications, both military and commercial. They must be willing to include in their plans and expenditures the people, time, and money needed to do the sort of reliability job that is required by the systems needs and specifications. An associate editor of *Electronic Industries* has commented, "For survival in the future, your company's integrity must be above reproach. In some of the new military specifications that are coming out, integrity of your company will be one of the evaluation points used when issuing contracts."

Motivation continually maintained is another major factor in obtaining reliability. One element which is effective in maintaining motivation is a sincere concern for reliability on the part of every manager in his dealings with his people, not only in formal "performance appraisals" but also in day-to-day conversations with them. Also, in his own decisions—in planning, organizing, and integrating, as well as measuring—"actions speak louder than words." Although each manager at each level is subject to his own motivation, i.e., his superior's attitude, each manager as well as each individual must strive to do his best to accomplish the desired reliability goal.

One method which has been found helpful in reliability motivation is to use movies developed for this purpose. The effects of such showings are of course temporary, and it is necessary for refresher treatments at recurring intervals. Another method of motivation is the use of short meetings of intimate-sized groups of people involved to discuss progress, records, and suggestions for maintaining and improving reliability. These meetings also must be held on a regular basis to maintain continued motivation.

Simplification of design is the third major factor in striving to meet the required reliability goals. The increasing complexity of systems in recent years has brought about the need for more and more equipments to work together effectively for the success of the overall mission. Since the fewer the number of parts, the less chance there is for failure, designs with fewer parts tend to be the most reliable.

Many a designer is tempted to demonstrate his cleverness by providing many special functions in the system he designs. These functions in many cases are used only very infrequently and may not be essential to the safe and satisfactory operation of the equipment. In every design review for each item of the system the following questions should be asked: "Is that circuit or that subassembly really necessary?" and "Could the system be made to work and to be more reliable without it?" If so, leave it out and simplify the design.

Reliability Check List

As a means of providing management with a convenient check list for reviewing the reliability position on each contract, R. H. Norris has prepared the reliability check list of Table 7.6-1, which has been most helpful in the preparation of the material for this chapter. Many of the items on the list have been discussed in the preceding sections. A few of the remaining will be considered in the material which follows.

Although many of the reliability check-list points are technically oriented from an engineering and/or manufacturing point of view, others related to education (19), utilization of specialized knowledge (20), utilization of experienced personnel (21), and auditing periodi-

Table 7.6-1. *Check List of What is Needed to Achieve Reliability Goals in Government Contracts*

1. Involvement of top management in support of reliability—essential.
2. Motivation of all individuals, continually maintained.
3. Simplification of design: "What is absent can't fail."
4. Standardization to reduce the number of varieties of component parts.
5. Production volume—in large numbers, uninterruptedly.
6. Resistance to design changes until their reliability is proved adequate.
7. Safety factors, conservatively chosen, based on evidence and statistical analysis.
8. Risk analysis—studies to allocate risk to subsystems.
9. Revision of failure-rate values when justified.
10. Tests to prove achievement of reliability goals.
11. Quality control: process control and cleanliness.
12. Information collection on failure incidents ("symptoms").
13. Investigation to diagnose the cause of the symptoms.
14. Initiation and execution of corrective action.
15. Comprehensive use of screening and burn-in of parts.
16. Use of redundancy when other procedures are insufficient.
17. Coordination between components of the organization.
18. Sophistication in knowledge and control of hazards.
19. Education: both technical and motivational.
20. Utilization of specialized knowledge both inside and outside the department.
21. Utilization of experienced personnel with good records of reliability achievement.
22. Records to make available the lessons of experience.
23. Use of knowledge by check lists and design reviews.
24. Application of these methods to vendors.
25. Auditing periodically of use of all these methods.

cally the use of good reliability methods (25) are primarily a responsibility of management.

Education. The need for a sophisticated approach to many of the factors contributing to failure is evident from the material in this chapter. A technical education program built around the work of the engineering organization and using one or more of the many textbooks on reliability now available can be an effective means for meeting some of the educational needs. Such a reliability course can be helpful to younger people new to the organization as well as to older people who may not have been educated to ideas of reliability risk and statistics.

Another kind of subject matter deserving emphasis in any useful education program for reliability engineering is the long-established "laws of nature." Frequently in a reliability course such material should be treated from the point of view of cutting across the traditional fields of college training. For example, exposing electrical and electronic engineers to the effects of heat transfer and vibration on the life of materials and exposing mechanical engineers to new computational principles and methods represent a few of the educational approaches that might be used. Modeling ideas may also be helpful in challenging engineers to new ways of looking at problems affecting reliability.

Utilization of Specialized Knowledge. The education program mentioned above is a slow, long-time method of solution to the need for better reliability information. Provision of a staff of specialists in each of the fields of knowledge requiring frequent consideration in reliability problems of the particular organization involved is a faster method of making available the greatly increasing body of information now at hand. Either the hiring of individuals to contribute their knowledge on a full-time basis where this can be justified, or the bringing in of consultants from private practice or the universities on a part-time basis, should be given consideration by management. In these times of rapid change, especially of information and equipment, it is important to have wise counsel available.

Utilization of Experienced Personnel. To achieve reliability, as in other matters, experience is a valuable asset. Personnel experienced in system synthesis and equipment or project direction with good reliability records are much more likely to provide reliable systems in the future than those who have had no such records or, worse, have had histories of failure in reliability matters. Even if the reliability requirements in the past have not been expressed in quantitative terms such as are now becoming required in contracts, other records, such as losses due to customers' complaints, may be resorted to as indicators of the ability to design reliable products and systems. Management should consider a man's reliability performance and attitudes in personnel selection for key posts.

Periodic Audit of Reliability Methods. Reliability objectives will receive continued attention from those involved only if management itself exercises eternal vigilance in this regard. One useful suggestion here is that some one individual in the department be assigned the duty to monitor periodically, or to audit, the effectiveness of the use of the adopted methods. Use of auditors is common practice to check on the adequacy of accounting; why not an auditor of the methods of promoting reliability? In the final analysis, however, management itself must carry the chief responsibility for reliability.

7.7 Conclusions

Reliability is an increasingly important aspect of the value of a system. It must be designed into the system on the basis of knowledge of the stress to be encountered and the strength of the system and its parts. From experimentally determined failure rates and failure mode characteristics, the systems engineer must understand the various possible failure mechanisms and minimize the likelihood of their occurrence. Consideration of environmental changes, tolerances, and effects of parameter changes and tests to verify the adequacy of the design are all part of the reliability effort. Design reviews and feedback of results from manufacture and operation help round out the emphasis on reliability by designers.

There is also an important quality-control effort required for obtaining a reliable system. Process control, cleanliness of the manufacturing operation, investigation of the causes of failure, and prompt and adequate corrective action are typical of the efforts required in the quality-control area. Thorough maintenance procedures properly adhered to are likewise a necessary part of the overall reliability effort.

Finally, management itself must be convinced of the importance of reliability and act in such a way as to instill in its people its feeling that reliability is essential and must be an integral part of the system's characteristics. Reliability can be obtained only with exceptional effort continually maintained.

8

Conclusion and Prologue

8.0 Introduction

The preceding chapters have provided a basis for organizing, formulating, structuring, and judging the performance and value of a system. They have also showed ways of evaluating the cost, time, and reliability of a system. These methods for visualizing and describing a system, its life history, and its different parts and phases, have utilized the systems engineering tools described in the earlier book of the same name. In many ways, one might be inclined to say that the subject of systems engineering has been covered and is closed.

However, realistically it is much more appropriate to take the point of view that the subject of systems engineering has merely been properly introduced; the tools and methods presented are adequate but will not be fully satisfactory for the increased number and complexity of the sytems of the future. Elsewhere in the real world, the seeds of change have already been sown in many areas—where systems are needed but do not exist, where systems exist but need to be improved, where improved materials, components, and equipments can make improved systems, where newly educated peoples are more receptive to new systems possibilities, to name but a few.[55, 63, 64]

The rapidly expanding population of the world is a driving function that is forcing a more extensive systematic approach to using the space, time, resources, and plants that we have. Newly understood methods of transmitting, storing, computing, regulating, controlling, sensing, displaying, and otherwise handling information are making it possible to satisfy increased human needs for more people with less physical effort by those involved. Social and economic systems, as well as physical systems for utilizing energy and making and forming material, are receiving increasing consideration but require even more attention.

The finite limitations on the world's space, natural resources, and human capabilities on the one hand, and man's semi-infinite ability to modify and alter the world on the other, provide a set of varying forces for change that place a premium on the need for better systems understanding. Man's ability to transport himself and other things more quickly, his increasing ability to develop new man-made and machine-made materials and techniques for handling them, and his ability to achieve better control and instrumentation of information—all these mean that we have the basis for the continued and varying changes with time that foster the need for systems engineering activity. The following few examples serve to indicate some of these changes.

Space. Until recent centuries, space was an obstacle to man. The Earth was too big for a man to travel around and explore in one lifetime. Now many tens of thousands of people have encircled the globe in a few days, and some individuals in less than 100 minutes. In the past there were too few people to accomplish a given task or to cope with natural phenomena; now population experts are concerned that there soon will be so many people that there will be no room for them on the face of the earth. However, by improved means of construction we can shelter many more people than are now being housed, we can grow and process much more food than we are now handling, and we can share facilities and use recreational means much more effectively than we have in the past.

Natural Resources. It took tens to millions of years to develop some of the natural resources which man has used up so quickly. Wood, coal, oil, and forests are typical of these. On the other hand, we have learned in the recent past, and are now learning at an increasing rate, to make substitute fuels, fibers, and materials, and to provide facilities for recreation and inspiration which can serve similar functions to those of the past. Also, our rate of making these new materials and facilities is comparable to, if not in excess of, our rate of consumption.

Human Capabilities. Although man's senses and other human faculties, such as being able to read and comprehend, have finite limits, our overall capability to sense, actuate, store, display, compute, transmit, and otherwise manipulate information has been greatly improved by instrumentation and control means. In centuries past, very little of the prior or contemporary culture was capable of being retained in a fashion comprehensible to many people; now with mass means for printing, storing, projecting, and retrieving, whole new vistas for the broader development of man and his intellect are opening up.

New ideas, new systems, and new social means for using these potentialities are required. Improved systems engineering is needed to make possible the realization of these opportunities for the future.

Summary of Tools and Methods 357

In this chapter, we will quickly summarize the systems engineering tools and methods contained in this and the preceding volume. Then a brief description will be given of a number of typical systems problems which exist and will continue to command the attention of engineers in the years ahead. It is systems problems such as these which provide a prologue to the future.

8.1 Summary of Tools and Methods

To a certain extent the terms systems engineering tools and systems engineering methods are somewhat complementary in nature and by no means unique. As we conclude this book on systems engineering methods, it is desirable that effort be made to place these methods in a matrix array and align them for comparison purposes with the systems engineering tools mentioned in the companion volume of the same title. Figure 8.1-1 shows such a matrix array.

In terms of the significance of the tools and methods upon one another, the array is purposely not filled in, for the detailed relationship may differ from system to system. In reality this matrix format serves to provide an organizational means for viewing the various aspects of the methods—namely, considerations of environments, organization, formulating and structuring, judging the value, considering cost, time, and reliability—and to place them in close proximity to the various tools of energy, materials, and information; modeling; computing; control; probability and statistics; signals and noise; optimization; and tolerances and errors which bear on these methods. Thus, although both methods and tools form coordinate systems within themselves, the location of the methods and tools in proximity provides a basis for starting a compilation of knowledge and information about a system which can be valuable in the early stages of comprehension and formulation.

The definition of a system and the statement as to how systems engineering is performed, which have been used in this book and the preceding volume, themselves provide a very useful check list with which the engineer starting work on a system is able to better appraise what his problem is. The definition, which describes the system as *an integrated whole although composed of diverse structures, parts, and subfunctions,* raises these questions: Just what is the total system being considered? What are these parts and how can they be partitioned, if you will, so that they can be handled separately in working on them and yet will operate together

	Environment for Methods	System Organization, Schedule, and Records	Formulating and Structuring	Judging the Value	Cost	Time	Reliability
Energy, materials, information							
Modeling							
Computing							
Control							
Probability and statistics							
Signals and noise							
Optimization							
Tolerances and errors							

Systems Engineering Methods (columns) / Systems Engineering Tools (rows)

Figure 8.1-1. Matrix array of systems engineering tools and methods.

Summary of Tools and Methods

smoothly? The decision of partitioning or not the system parts is itself a subject over which the system designers may have some influence and should be considered as a variable in cases where this is possible.

The system, by definition, *has a number of objectives*. What are those objectives? What is the relative importance of them? These objectives are mentioned as being different from system to system. To what extent need they be different? To what extent are they similar to those of other systems? If they are similar, to what extent can this be capitalized upon in the consideration of the system?

The matter of *compatibility* is an important factor for the system designer to keep in mind. As has been indicated, the making of a system is of itself a system. Engineering, manufacturing, purchasing, etc., should be performed with the thought of retaining an overall simplicity to "the system for making the system" by keeping a measure of compatibility of these features from one system to another.

The term *optimization* is an important part of the definition of the systems engineering method and serves to emphasize the facts that in all of our endeavors with human beings and with the use of resources there is in general a choice of ways of doing things and that certain ways may be more attractive and suitable than others. On this basis, these choices of alternatives must be considered; and the use of optimization, whether it be formal and mathematical, or heuristic and intitutive, is optional to some degree to the person performing the optimization. But in any event this definition of the system and the method of performing systems engineering does of itself provide a useful means for indicating the way in which systems can and should be viewed in their overall concepts.

The concepts which were spelled out in the earlier book, *Systems Engineering Tools*,[49] serve as a foundation in the matter of systems methods. These concepts, as tabulated in Table 8.1-1, can be helpful in developing or improving a system. They emphasize a set of considerations of change which involves time, both history of the past and prediction for the future. The alternative ways of accomplishing things present a challenge to consider a number of the methods possible for achieving the desired system objectives. The alternative bases for judging the value of the system provide a framework within which the alternative methods may be evaluated and compared. Consideration of the system environment forces one to look outside the system being engineered and to examine the more comprehensive system of which the present system is a part. And, finally, the potentialities

Table 8.1-1. *Systems Engineering Concepts*

1. Any system is continually changing with time.
2. There are alternative ways of accomplishing things.
3. Commonly accepted bases exist for judging the value of a system.
4. Each system has its own environment.
5. Computational and experimental techniques exist as alternatives to actual construction.

of computation and experimentation provide alternatives to actually building in life size and to full scale the system being considered. Thus, these systems engineering concepts not only are useful in understanding systems in general but also can be most valuable in terms of the way in which each particular system is conceived, designed, and implemented.

The approach used in solving systems engineering problems, as shown in Table 8.1-2, includes the processes of formulation, structuring and synthesis, design and construction, measurement, and evaluation. These five portions of the overall method serve to provide a framework in time on which the system life may be conceived and structured. Thus they can provide an indication of the different phases of the systems engineering job which must be accomplished. In this way one can at the outset of the job block out the general items of work to be done and form a basis for synthesis by means of which the individual items of work can be accomplished.

The brief summaries above provide an indication of many of the decisions which must be faced in performing our systems engineering job. They serve to point out the need for defining the space-, time-, and applications-oriented environment in which the systems will operate, and the money, people, and resource environment in which the systems engineering and construction must be performed.

From this brief summary of these systems engineering tools and methods, what is the message that comes through? The message is that the systems approach is a broad approach to handling a problem,

Table 8.1-2. *Approach Used in Solving Systems Engineering Problems*

1. Formulate the problem.
2. Synthesize the system to perform the requirements.
3. Find ways to make the system synthesized.
4. Measure what has been done and compare with the objectives.
5. Use error from comparison to refine previous steps.

not simply meeting the apparent and obvious needs but stressing consideration of the requirements over the entire span of system life. The systems approach is not a unique one, but it represents an organized attempt at looking at the objectives, methods, and results in their entirety rather than merely performing a limited set of activities which might appear to be based on a restricted, subsystem point of view. The systems approach stresses variety, not uniformity, in the individual systems objectives. The systems approach stresses uniformity, not variety, in the modules or elements of hardware making up the system.

The systems approach includes the subjective considerations of value judgment, cost determination, and time significance. Thus, the systems approach endeavors to perform the interface function of meeting the requirements and desires of the people who will use the system and, at the same time, keeping in mind the needs and capabilities of the people who are making the system. Systems engineering serves the useful function of providing a middle ground where the supply and the demand in technical, as well as functional and economical, considerations are explored and reconciled.

8.2 Typical Systems Opportunities for the Future

Armed with a sound understanding of systems engineering tools and methods and having available much capability in both hardware and software, we can profitably look at some of the challenges and opportunities that exist for the future. In the section which follows, consideration will be given to several areas, such as information systems, energy systems, material systems, community growth and development systems, and transportation systems. In most cases these systems areas are by no means new subjects; however, by virtue or new needs and objectives, new environmental conditions in the state of the art and in customer attitudes, and our improved willingness to try new things, we can be planning and working toward new systems. And it seems apparent that these new systems themselves will probably be merely the prologues to better systems of the more distant future.

Information Systems

Heralded widely as the most recent of the revolutions to affect markedly our way of life on earth, the information revolution follows the industrial revolution and the revolution caused by automation and automatic control. Certainly man's ability to generate, transmit, store, retrieve, manipulate, and display information and data per unit of time has increased many orders of magnitude, and the cost per unit operation has decreased markedly. Information systems problems for the immediate future range from the more efficient handling of some of the computer hardware systems we have at present, through the development of more efficient equipment and methods for data storage, to the more effective organization and ready availability of business management systems for industry or government. Although these examples are merely representative, let us consider a few of them in some more detail.

Remote-Access, Time-Shared, Digital Computer Systems. In recent years the advent of digital computers with remote-access, time-sharing capabilities has brought into interdependence a number of users for a central computer facility. Figure 8.2-1 shows a central computer connected by means of a data network to many remote individual users in addition to its local load of computation needs. By means of a telephone or other communication means, the individual remote subscriber can signal the central computer of his desire to use it, and by means of a priority system in the computer for selecting which user shall have how much time (measured in milliseconds or less), the computer shares its time in the concurrent solution of a number

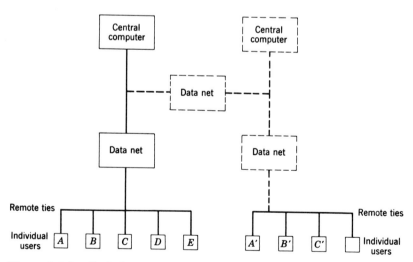

Figure 8.2-1. Central computer connected by data nets to remote users and other central computers.

of users' problems. Because these many solutions are being processed by the computer on an essentially real-time basis, the term real-time, multiprocessing is used to describe this sort of digital computer system operation.

The various individual users that have availability to the computer may range from tens to hundreds in number. They may share the current time usage of the central computer and associated peripheral equipment as well as the longer time memory and other data storage. The broad framework of organization, including memory, control, and associated equipment, represents one set of systems questions to be considered. The size and number of data nets, computers, and individual-user terminals, shown by the dotted lines in Figure 8.2-1, as well as the priorities and charges assigned to the individual subscribers, are additional parameters which require solution by systems engineers. The answers to these questions are ones which will have significance over a single period of time, but as equipment techniques, customer demands, and other factors change, the systems solutions possible will doubtless become different in the future.

More Extensive Memory Systems. A subset of the problem noted above is the need for more extensive memory systems. Actually, there is a cost, size, and speed trade-off in memories; some, such as core memories, operate in micro- and nanoseconds, whereas others, such as punch cards and tape, are much slower. All provide the somewhat similar function of data storage to a first approximation. However, by judicious design it is possible to utilize some of the fast core memory to take into account the amounts and locations of the memories in other forms, and this capability should be brought to bear more effectively in overall integrated data storage systems. Figure 8.2-2 indicates schematically how these individual memory means might interact with one another. The nature of the hardware and software means for accomplishing the different levels of memory functions and the requirements for each are dependent on various forms of natural phenomena, as well as on skill of the design of the more extensive memory systems for the future. The advent of new fast-access, high-density storage means can precipitate new opportunities in the information systems field.

Storage and Retrieval Systems. Another class of information systems problems involves the so-called storage and retrieval systems, which will be used more extensively in the future for specialized libraries serving various technologies and functional interests. For example, the need to retrieve previously stored information on patents,

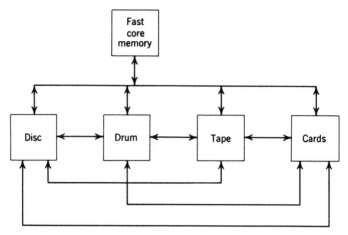

Figure 8.2-2. Interaction of various types of memories into an integrated data-storage system.

medical diagnoses, chemical data, and business operations exists for different users and in varying degrees. The method for organizing the storage of information in any one of these fields is itself a problem of considerable magnitude for the practitioners in that field. It is apparent that, as we change our understanding of the universe and those parts of it which are significant to us, the best organization for a storage system to record this information will indeed change with time. Furthermore, apart from the requirements that information be stored and retrieved in any one special interest area, such as patents or chemicals, there is the generic problem of selecting sound basic methods by which these large storage and retrieval systems will function, even though the means used for storage and retrieval may vary somewhat for different applications. For any one area of technical interest, there may be many features of the storage and retrieval systems which are similar to those for other technical areas. Thus, there are alternative approaches from the systems point of view: one which stresses the applications aspect of information storage and retrieval and which is user-oriented, and another which is hardware- or software-oriented and emphasizes the mechanistic, equipment, and programing software aspects of storage and retrieval.

Inventory and Control in Management Systems. With the control of a particular production process, as, for example, the manufacture of steel or the making of electrical power or of chemicals, there is associated an information system which keeps track of the needs of

Typical Systems Opportunities for the Future

the various customers as reflected in their orders and of the supplies of the various materials, manpower, equipment, and facility resources necessary to make and assemble the desired products. Thus, as shown in Figure 8.2-3, there is an overall order-processing information control system for the process system. This manufacturing inventory and control system coordinates and controls the orders for a product or a number of products with the individual materials and items which go to make and assemble the overall product. In some cases, these orders are for equipments which are standard in nature, and it is possible to stockpile completed units, such as light bulbs, clocks, radios, or motors. On the other hand, the equipment desired may be very specialized, as, for example, a high-energy particle accelerator, a steam-turbine generator set, or an interplanetary communication system, in which event the likelihood of it being available on the shelf is very small. In each case, it is necessary for the business manager or the project manager involved to keep a balance between the needs of his organization, which is making the products and equipment, and the time schedule and requirements of the person who is ordering or requesting them.

Business Management Information System. Another form of information system is the more comprehensive business management system. For example, if the overall business to be managed can be considered as composed of a number of parts, such as manufacturing, engineering, finance and marketing, and their associated subfunctions, there exists a need for maintaining up-to-date knowledge of the relative position of each of these functions with regard to their basic

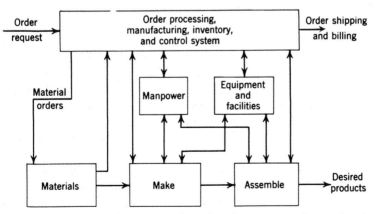

Figure 8.2-3. Order processing, manufacturing, inventory, and control system.

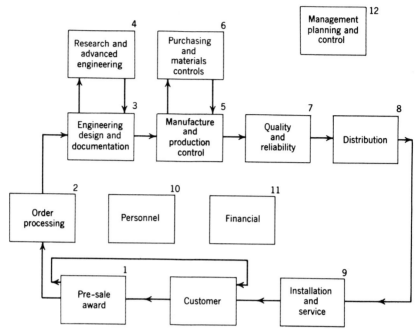

Figure 8.2-4. Generic business model showing function subsystems.

estimates, their current position. Furthermore, the needs for keeping track of the business environment, of the competitor's position, of changes in actual purchases of the market versus anticipated purchases, all require a great deal of information. The organization of such an information system is at one time specialized as it relates to a particular business and at the same time can be somewhat generic in relation to the fact that, although businesses may have different products, they often have somewhat common methods of handling their administrative organization. Figure 8.2-4 shows a block diagram of a possible generic business model in which are twelve functional parts. The very existence of business schools at the collegiate and university levels is evidence that there are generic problems common to all businesses. The fact that certain businesses are engaged in one industry and others in different ones indicates that the nature of the information in one business management system may be unlike that in another. However, many of the functions may be common to several businesses. Here again we have this need for a systems engineering approach to correlate the capabilities of the equipments

Typical Systems Opportunities for the Future 367

to do jobs with the requirements of the individuals or groups which have jobs to be done.

Another aspect of a business management information system presenting opportunities for the future can be described as one embracing data communication networks that are more geographically widespread and more selective, business-wise. Thus, as individual businesses become geographically decentralized, it is highly desirable that they be integrated to one another by means of data communication networks. In the book *Systems Engineering Tools*, such a data communication network is described. Obviously, the needs for a network of this kind will change with the volume of load to be handled and with the extent of new equipment developments in the communication field itself. As an example, the transition from submarine cable to wide-band television communication satellite may alter data rate and volume capabilities and provide opportunities for new and different information systems.

On-Line Test Data-Processing Systems. Still another form of information system which can be extremely meaningful and valuable is used for the on-line test data reduction of experimental programs. In many cases, currently, tests are made either on a research, developmental, or perhaps manufacturing basis on large equipments in which the lag between the time the test is made and the time the results are known is significant. Figure 8.2-5 illustrates in a block diagram fashion how the test control and data processing, the many sensors,

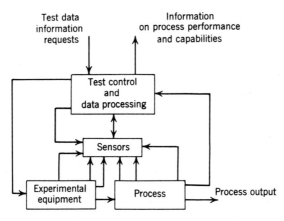

Figure 8.2-5. Test control and data processing for on-line control of an experimental process control system.

and the experimental equipment and process itself form another more comprehensive on-line system.

In many cases the engineering data and charts that are sought may take weeks or months to obtain. Meanwhile, a question exists in the minds of the people performing the tests as to whether or not the equipment and process are satisfactory for the purpose intended. If it is, the equipment should be removed from the test area, or perhaps other experiments should be run, depending on the nature of the results. If the test is not successful, then changes may be required at once. During the period which elapses from the time of performing the tests to the time when knowledge is available of the results, the choices available to the person having to make the decisions may involve many people, much equipment, and a large sum of money. By means of on-line processing for experimental data reduction, it can be possible to obtain much greater economy of facilities, people, and resources. The cost involved in the on-line test data reduction system may be by no means small, but on the other hand the corresponding savings to be realized in the form of more efficient operation and lighter investment of capital goods can likewise be very significant.

From the above brief description of several new or improved information systems which have been suggested for the future, it is apparent that there are a number of worthwhile and challenging opportunities for systems engineering in the information systems field.

Energy Systems

In the area of energy systems, many problems of a systems nature are looking for better solution in the future. This is not to imply that solutions are not being obtained currently or that satisfactory solutions do not exist. Rather, in the context of things changing with time, the needs for improved energy systems likewise exist and require proper attention. With the number of people in the world increasing rapidly, and the energy usage per person also rising, future energy needs are monumental.[45]

Broader Interconnection of Power Systems. The first of these problems is associated with a broader interconnection of power systems on a more integrated basis. As the use of electrical power becomes more extensive and as the operating voltage increases in magnitude, it is desirable to interconnect power systems to achieve the economies of operation which are possible with this broader basis for the generation and use of power. Thus, as indicated in connection with the

Typical Systems Opportunities for the Future 369

growth of systems in Chapter 3, it is possible and desirable for isolated systems to be combined and to be operated as an integrated whole. Although, empirically, energy systems have been growing bigger and bigger, the rules and methods for operating such systems with the increasing reliability necessary will demand the attention of systems engineers operating at a higher level of systems integration. The increasing severity of power failures on individual electrical systems and the repercussions of such failures on the still broader systems of which these individual systems are parts have focused attention on the need for greater understanding of broadly interconnected power systems. Although improvements to existing interconnection methods and practices are currently available, still more advanced systems methods may be required in the future.

Longer-Distance Transmission of Large Blocks of Power. Another energy system problem of considerable importance which is receiving greater attention is associated with the longer-distance transmission of large blocks of electrical power. For example, in the case of the United States and Canada, large sources of hydropower exist in Canada where the need for such power in reasonably adjacent geographic areas is too small to warrant the development of these potentially huge sources. On the other hand, in the United States, where there exist large electrical loads corresponding to heavy concentrations of population it would be desirable to have very much larger blocks of power at cheaper rates than are currently being developed locally. Through the use of high-voltage electric power ties, longer-distance transmissions of large blocks of power can provide economically a common system consisting of the large energy sources and the large energy demands. The problems associated with bringing together the energy systems of generation and use are many, and range from the fundamental difficulties associated with handling large blocks of power to the complexities of the corresponding information and control systems necessary to time the arrival, initiation, and use of these large energy systems. Figure 8.2-6 shows some of the geographical areas involved in North America; and many other similar instances for long-distance transmission of power exist elsewhere in the world and also require systems attention.

Jet and Rocket Engines. Another form of energy system which is receiving and will require additional effort is that associated with the larger jet engine and rocket propulsion systems for aeronautic and space applications. The development of extensive aircraft facilities in the form of airports and communication systems and the desire

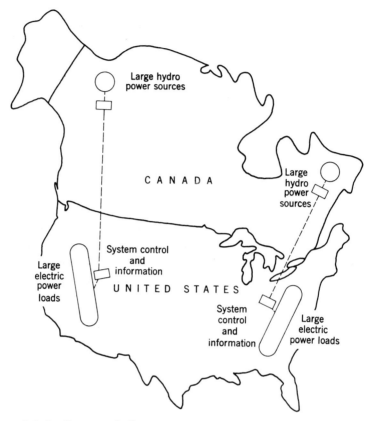

Figure 8.2-6. Systems challenges afforded by the need for interconnecting large hydropower sources with distance loads.

of more people in the world to be able to travel to foreign lands have provided a greatly increased market for transportation methods, such as aircraft, capable of moving larger loads of people and equipment over long distances rapidly. Typical of this expansion is the work currently taking place on large military transport carriers and supersonic transports, as well as large passenger subsonic transports. Here again, the requirements of the energy systems are many and interdisciplinary in character. In addition to the basic need of being able to develop the inherent mechanical power efficiently in these larger-size units, it is also necessary to provide for extensive controls to initiate, use, and shut down such engines under many different conditions. Furthermore, in many cases it is essential to control many intermediate portions of these engines so as to provide an optimum

Typical Systems Opportunities for the Future

operation for the overall engine and a number of its component parts. Fuel systems, actuating systems, and temperature-control systems, which are necessary parts of smaller engines, become even more vital as the engine size increases.

In the area of rocket propulsion systems for space applications, again there are needs which far surpass the present status of such systems. Greater efficiency, more flexibility in starting and stopping the engines, and better fuel utilization are typical of the energy systems problems which exist and which will require attention in the future. In addition to the need for larger jet and rocket engines, specialized power sources for use in smaller-sized ratings will doubtless require additional systems attention. Whereas large systems can frequently justify the cost and effort devoted to such problems as heat transfer, mechanical strength, and similar significant matters in the smaller systems much more ingenuity and attention may be required than can economically be brought to bear.

Energy Systems Using Alternative Power Sources. Another aspect of energy systems requiring attention for the future is the combination of alternative energy sources for the most economical system operation. In many cases the expenditure of moneys for fuel and facilities to provide energy generation is a significant part of the total expense of the operation of the energy system, for example, those using hydropower sources, steam power sources, fossil fuel sources, or pumped storage sources. Although currently much attention has been given to very large-scale integrated operation of systems from a power-systems-economics points of view, there are many smaller industrial and other systems which could benefit from such treatment. Furthermore, as possible new developments in exploiting older energy sources take place with time, the criteria for decision as to when one or the other or a combination of several energy sources will be used may change. It appears likely that additional emphasis on systems using alternate energy sources will be required in the future.

Materials Systems

Radical changes have taken place in the materials with which man lives and does his work. Future improvements in materials systems are likely to be obtained in a number of different fashions. There will be modified methods of obtaining and preparing the materials we already know and use. Systems will be devised for obtaining and processing new materials which are yet to be developed. There will be new systems for the fabrication of devices and for the construction of edifices from new and old materials. Plastics and microelectronics are just two important materials which are only now beginning to be exploited by modern man. There are and will be many others.

Improved Methods of Making Existing Materials. The known natural supplies of such existing materials as iron, steel, chemicals, and petroleum are continually being expended, and new explorations have not always been able to locate additional sources of as high quality or availability as the ones in current use. Fortunately, there is an increasing amount of previously processed material which may have some value if suitably reprocessed and made available. In many cases new methods and systems of processing will be required to obtain materials of usable grade from less pure basic raw materials. Sintering and concentrating systems for ore enrichment have been worked out. Quicker methods of iron reduction, such as those employing the basic oxygen furnace (BOF) process, have made possible much more rapid change of iron into steel. Likewise, in chemical and petroleum processing systems new computer controls have been bringing about products of better quality with less cost and in a shorter time. Emphasis in these areas will involve better process knowledge, more efficient systems organization, and more extensive control. Because many such basic systems involve large amounts of energy to perform the materials conversion processes, frequently systems for exploiting both the materials and the energy portions of the undertaking will have to be developed.

Systems for Making New Materials. The increasing demand for new materials having specific combinations of properties, such as low weight/volume, high strength/weight, high volume/cost, low conductivity/volume, or high conductivity/volume, and an increasing knowledge of and capability for handling chemical operations and processes have brought about many new materials. Polymers, silicones, fibers, and films are representative of such products. The methods for obtaining the necessary raw materials for making these new materials, the operations for processing and modifying them as the various steps in their manufacture are performed, and the shaping and storing of them before their ultimate use present a number of interesting and challenging systems problems. Various alternative ways for accomplishing the results and alternative value judgments may exist for these different materials, so that creative systems engineers in such fields will be stimulated to conceive and put into practice new systems for making an increasing number of new materials in the future.

Improved Construction Techniques. The development of the mass market for materials and products, the increasing emphasis on modular construction, and the lower cost for using mass-produced prefabricated parts have introduced a number of new construction techniques in the recent past. Furthermore, the availability of larger, more specialized, and more sophisticated machinery has made it possible for new systems to help man in novel and improved methods of construction. Large earth movers and shovels, graders and plows, and versatile hoists and cranes are being used more extensively in the construction trade, and improvements are likely for the future.

The concept of using a few basic construction shapes and modules for buildings and placing them in various combinations has opened new possibilities for putting together the many new apartments and other dwellings required for the doubled world population predicted to exist a mere forty years from now. Just as we can now place a prefabricated wall into position rather than laying the wall brick by brick, so new construction systems for the future will be engineered for greater speed and economy. The mass-produced factory-built assemblies will require new systems for their construction; installation and use of these assemblies will in turn necessitate other new systems.

Automated Manufacture, Design, and Modular Construction. Hand in hand with the development of new materials and processes are the allied ways of making the resultant end products to be used by the consumers of the various devices and equipments. For example, microelectronics with its greatly decreased size and weight of material has brought about new methods of fabricating many electrical and electronic devices. What systems of men, machines, and equipment will be "best" to exploit microelectronics for the future? Certainly the manual assembly lines of the past may not prove to be the answer.

Automation, often in the past criticized for taking jobs away from man, may be more eagerly sought as a means of helping the people of the world to realize their increasing desires for growing numbers of material things. What systems for using automatic means for production can most effectively be employed to achieve these material aspirations? Surely good systems engineering will be required here both for basic design and for means of manufacture.

The automatic programing of tools (APT) system of machining represents one example of how production methods have been changed in the past. Design methods using man and computer, such as

SKETCHPAD and similar coordinated information systems, will doubtless be expanded in the future. Automatic design optimization methods (ADOPT, AID, and others) should likewise point the way to improved design for the systems of tomorrow.

The concept of modular construction in which various combinations of relatively small but versatile modules can build up large and complicated systems has been demonstrated vividly in the recent past. Modularity of both hardware and software is a driving force for and a product of systems engineering. Modular construction, modular test, and modular programs are merely a few examples which systems engineering will influence in the future.

Community Development

The pre-twentieth century concept of a static society in which the rate of change of population was small and the rate of obsolesence of buildings and communities was low has been replaced by a more dynamic representation. Large sections of cities which are barely 100 years old are being replaced by newer buildings and constructions. Entire new communities of 10,000 to 50,000 people are being planned and built as suburbs of existing cities. Where the limitations of the past are less severe, new cities of 200,000 to 500,000 are being planned as complete new metropolitan areas. Furthermore, in both the existing and new communities, increasing attention is being given to providing more effectively both hard services, such as water, sewers, and transportation, and soft services, such as health, education, entertainment, and welfare. How best can the solution of these community problems be handled in the future from a broader systems point of view?[57]

Rebuilding Delapidated City Areas. During certain periods of rapid development in many cities there were built homes, shops, factories, and streets which were not planned in terms of the long-run needs of these communities. Changes with time have provided new needs and new alternatives to meet these needs. The result is frequently that these areas, although desirable in the sense of now being centrally located, are problems in terms of representing a financial drain to the community and a deterrent to its growth and health. A number of methods of bringing about desirable changes have been tried with varying degrees of success. Since large amounts of money and resources and large numbers of people may be involved in these changes, even if only because of inconvenience during the transition period, the application of systematic methods of accomplishing alterations in older communities will receive increasing attention in the future. Typical of the problems involved are the technical difficulties of physical destruction and reconstruction and the social and economic upheaval required to move people and restructure the new environment. Interdisciplinary approaches are needed, and a new skill and understanding will be demanded.

Building Additions to Existing Cities. The movement of large numbers of people from farms to large cities has been accompanied by a growth of suburban areas around the periphery of these cities. In many cases it has been found financially desirable to provide suburban developments of both domestic and commercial character which are of questionable compatibility with the long-range interests of the overall urban area. In some cases, the outlying areas are not even in the same state, although there is little doubt that a common regional interest exists between the city and its suburbs regardless of local political boundaries involved. The efficacy of handling joint facilities, such as roads, water, sewers, parks, libraries, recreation areas, and shopping districts, in the long run will profit from a sound systems approach. Fortunately community planning is receiving more attention from government, industry, and the universities, and stronger systems efforts in this aspect of community development will doubtless be made in the years ahead.

Development of Whole New Cities. In the United States as well as abroad, attention is being given to the possibility of planning entirely new large cities to help accommodate the greatly expanded population of the future. In the past such capital cities as Washington, D.C., and Brazilia were planned and built in the wilderness. More attention to other new cities will doubtless be given in the future. Utilizing such natural features as climate, resources, geographic location, and power, it is necessary to combine economic, cultural, physical, and other environmental features in a solid systems evaluation of such new cities. Fundamental decisions regarding utilities, such as power, transportation, water, and sewers; industry, such as factories and warehouses; residential and shopping centers; schools and government buildings should benefit from an overall systems approach. Although the questions involved are by no means wholly engineering, there are many problems of a technical nature which require solutions over and above the financial and political ones which frequently command the major immediate attention.

Improved Water, Sewerage, and Pollution-Control Systems. The demand for communal services has grown rapidly in the past few decades to the point where such commonplace functions as water supply, sewage disposal, and air-pollution control are becoming major problems for communities and cities. Water shortages in recent years have emphasized the need for a systems approach to obtaining water, storing it, and delivering it to the user at a price and with a reliability compatible with his requirements and willingness to pay.

The handling of sewage and waste by cities and other communities was once a factor of hardly any consequence. In recent years, the cost to the community involved or to its neighbors has greatly increased and may be very large and significant. Also, in underdeveloped areas, public health considerations may soon be realized to involve heavy expense or loss to the communities affected. Sound systems efforts at balancing engineering costs and social losses may show the justification for increased emphasis on the sewerage problem both at home and abroad.

Air and water pollution from combustion products as well as from nuclear contamination are likewise being shown to be real or potential sources of economic loss to many present-day cities, states, and nations. Because the solutions to these problems are frequently both technical and economic, the application of systems engineering techniques is both attractive and essential. More studies of this sort are being undertaken, and the possibility for future systems activity in these areas is great.

Increased Efficiency of Community Services. Police and fire protection, handling of welfare for needy citizens, efficient provision for shopping and parking, availability of hospitals, libraries, parks, and schools are typical of the community functions which today's citizen may reasonably expect. Whether these services are state or individually supported financially, the means for providing them frequently require sound engineering understanding of the objectives and methods involved. Improved data-handling systems may be effective; regional development studies and construction may be needed; new constructive approaches to overcome the problems of relief from one generation to the next may be required. Here again a combination of technical, social, economic, and political considerations may be essential to meet the particular needs of the entire system involved.

Transportation Systems

The availability of more and faster vehicles to move more people further has placed new requirements on transportation systems. Streets, roads, and highways are now being built to new standards of safety, lighting, and speed. In developing countries where the transition from foot travel to jet travel will come in a decade, the whole gamut of transportation possibilities must be viewed from technical, economic, and social considerations. Also, the integration of land, sea, and air systems into a more convenient overall means for travel by the many different kinds of individuals and freight involved will be even more important in the future. The use of systems engineering can help the various governmental and private agencies participating to make sound technical, economic and social decisions on these matters.

Restructuring of Streets, Roads, and Highways in Developed Countries. In many cities and regions, the transportation systems in existence have developed over long periods from days when space, time, and economic values were based upon different considerations from those now prevailing. Means for modifying some of these old streets, roads, and rights of ways should be evaluated, using sound engineering methods, to see whether overall improvements are possible and justified in the light of present and future situations. The problems of appropriately retiring from service transportation means which have outlived their period of usefulness perhaps may receive increasing systems engineering attention. Likewise the question of bringing into more effective use the transportation routes and methods that should be made available for the future is one which systems engineers could profitably investigate.

Establishing Transportation Base for Developing Countries. In the light of the different needs for many of the developing countries, and the cost and availability of alternative methods of transportation, systems engineering studies will be employed in increasing number in the future to try to provide the most favorable transportation mix for the country involved. The system judgment standards of time, cost, performance, reliability, and maintainability can be employed to advantage for the particular parameters and objectives present. Perhaps some of the solutions to the transportation problem which have developed in a historical sense may not be warranted in today's world of the airplane, truck, and automobile.

Integrating Land, Sea, and Air Transportation. With some of our transportation systems moving at supersonic speeds, some at subsonic speeds, and others under the control of red and green lights, it would appear that benefit could be derived from looking technically at the interconnection means from system to system. Engineering investigations viewed in terms of the user's requirements in addition to the technical capabilities of the transportation media may prove to be increasingly worthwhile. Regional transportation needs in the light of today's and tomorrow's requirements may show the effectiveness of realignment of political boundaries established during the days of the horse and buggy and the once-an-hour ferry. These problems will not be solely technical, and they are doubtless not going to be simple. However, their solutions will probably not become any easier or cheaper, as in the future more people cause greater inconvenience and heavier expense. In the systems time sense, it probably pays to tackle such integration now, that is, as soon as one can reasonably understand what should best be done.

8.3 Conclusion

The scientific and technological achievements of the past and present and the ever-increasing material and spirtual desires of the rising world population are providing driving forces for change. Systems engineers, working with other technical people as well as those trained in social, economic, political, medical, and other fields, have an important and useful opportunity to help shape a better world. Let us get on with the job.

Bibliography

1942-1956

1. "American Standard Definitions of Electrical Terms," sponsored by American Institute of Electrical Engineers, (now IEEE), New York, 1942.
2. "What Price Speed?" T. J. Von Karman and G. Gabrielli, *Mech. Eng.*, **72**, No. 10, 776–781, October 1950.
3. "Engineering Economy," American Telephone and Telegraph Co., New York, 1952.
4. *Demand Analysis*, Herman Wold and Lars Jureen, Wiley, New York, 1953.
5. "Terms of Interest in the Study of Reliability," C. R. Knight, E. R. Jervis, and G. R. Herd, ARINC Monograph No. 2, Aeronautical Radio, Inc. Washington, May 1955.
6. "Handbook of Preferred Circuits, Navy Aeronautical Electronic Equipment, NAVAER 16-1-519," Bureau of Aeronautics, Department of the Navy, September 1955.
7. "Reliability Stress Analysis for Electronic Equipment, IR 1100," Radio Corporation of America; also same title, PB 131678, U.S. Department of Commerce, Office of Technical Services, Washington, D.C., November 1956.
8. "Engineering Studies of Economy," Long Island Lighting Co., about 1956.

1957-1959

9. *Introduction to Operations Research*, C. W. Churchman, R. L. Ackoff, and E. L. Arnoff, Wiley, New York, 1957.
10. "AGREE (Advisory Group on Reliability of Electronic Equipment), "Reliability of Military Electronic Equipment," Office of Assistant Secretary of Defense—Research and Engineering, Washington 25, D.C., 1957.
11. Topology of Switching Elements versus Reliability, J. P. Lipp, *IRE Trans. Reliability Quality Control PGRQC-10*, June 1957.
12. *Games and Decisions*, R. D. Luce and Howard Raiffa, Wiley, New York, 1958.
13. "Reliability Improvement Through Redundancy at Various System Levels," B. J. Flehinger, *IBM J. Res. Develop.*, **2**, 148–158, April 1958.
14. "PERT Summary Report, Phase 1," Special Projects Office, Bureau of Ordnance, Department of the Navy, July 1958.
15. *Servomechanisms and Regulating Systems Design*, Vol. 1, 2nd ed., H. Chestnut and R. W. Mayer, Wiley, New York, 1959.
16. *Systems Engineering*, H. H. Goode and R. E. Machol, McGraw-Hill, New York, 1959.
17. *Economic Control of Interconnected Systems*, L. K. Kirchmayer, Wiley, New York, 1959.
18. "Analogue Computing Applied to Plant and Process Economic Estimation," D. W. Gillings, Joint Symposium on Instrumentation and Computation in

Process Development and Plant Design, published by the Institute of Chemical Engineers, London, May 1959, pp. 121–127.
19. "Economic Comparison of Alternate Plans," Edison Electric Institute Electrical System and Equipment Committee, Report of System Planning Subcommittee, T. W. Schroeder, Chairman, May 1959.
20. "Integration of Systems Engineering with Component Development," J. A. Morton, *Elec. Mfg.*, **64**, 85–92, August 1959.

1960-1963

21. *Operations Research and Systems Engineering*, C. D. Flagle, W. H. Huggins, and R. H. Roy, eds., Johns Hopkins Press, Baltimore, Md. 1960.
22. *The Control of Multivariable Systems*, M. D. Mesarovic, M.I.T. Press and Wiley, New York, 1960.
23. *Principles of Engineering Economy*, 4th ed., E. L. Grant and W. G. Ireson, Ronald, New York, 1960.
24. *Dynamic Programming and Markov Processes*, R. A. Howard, M.I.T. Press, Cambridge, Mass., 1960.
25. *Systems: Research and Design*, D. P. Eckman, ed. Wiley, New York, 1961.
26. *Industrial Dynamics*, J. W. Forrester, M.I.T. Press, Cambridge, Mass., 1961.
27. "Reliability and Longevity Requirements, Electronic Equipment, General Specifications for," *USAF Standard MIL-R-26667A*, 1961.
28. "Standardization of Electronic Test Equipment," D. B. Dobson and L. L. Wolff, *Elec. Eng.*, 60–67, January 1961.
29. "Standardization of Electronic Instrumentation and Control Systems," J. G. Nish, *Elec. Eng.*, 129–135, February 1961.
30. "Quantitative Approach Gets Reliability Results," James Holahan, *Space Aeronautics*, **35**, No 4, 122–127, April 1961.
31. "Cost Models for Control Systems Engineering," Harold Chestnut, *AIEE CP 61-707*, May 1961.
32. *The Design of Engineering Systems*, W. Gosling, Wiley, New York, 1962.
33. "PERT, A New Management Planning and Control Technique," Jerome W. Blood, ed., American Management Association, Report No. T-74, 1962.
34. *A Methodology for Systems Engineering*, A. D. Hall, Van Nostrand, Princeton, N.J., 1962.
35. "Analysis and Synthesis of Dynamic Performance of Industrial Organizations—The Application of Feedback Control Techniques to Organizational Systems," R. B. Wilcox, *IRE Trans. Auto. Control*, **AC-3**, No. 2, March 1962.
36. "System Analysis Procedures for System Definition," Ballistic Systems Division, Air Force System Command, *USAF, BSD Exhibit 62-101*, June 1, 1961.
37. "Evaluation and Reliability," S. W. Herwald, *Elec. Eng.*, 514–516, August 1962.
38. *Systems Reliability Engineering*, G. H. Sandler, Prentice-Hall, Englewood Cliffs, N.J., 1963.
39. *American Standard Terminology for Automatic Control*, *ASA C 85.1* 1963, American Society of Mechanical Engineers, New York.

40. "PERT Cost System Description Manual," Vol. III, Air Force Systems Command, December 1963.

1964-1967

41. *Views on General Systems Theory*, M. D. Mesarovic, Wiley, New York, 1964.
42. *Maintainability*, A. S. Goldman and T. B. Slattery, Wiley, New York, 1964.
43. "System Engineering Management Procedures," *Air Force Systems Command Manual (AFSCM 375-5)* USAF, February 1964.
44. "Learning Curve Approach to Reliability," J. T. Duane, *IEEE Trans. Aerospace*, **2**, No. 2, 563–66, April 1964.
45. "Growth of Energy Consumption and National Income Throughout the World," Fremont Felix, *IEEE Spectrum*, **1**, No. 7, 81–102, July 1964.
46. "Common Foundations Underlying Engineering and Management," J. W. Forrester, *IEEE Spectrum*, **1**, No. 9, 66–77, September 1964.
47. "Systems Engineering," H. A. Affel, Jr., *International Science and Technology*, 18–26, November 1964.
48. "Understanding the Engineering Design Process, J. Morley English, *J. Indus. Eng.*, **15**, No. 6, 291–296, November–December, 1964.
49. *Systems Engineering Tools*, Harold Chestnut, Wiley, New York, 1965.
50. Weapon System Effectiveness Industry Advisory Committee, (WESIAC), *Final Report of Task Group I, Requirements Methodology AD 458453*, January 1965.
51. Weapon System Effectiveness Industry Advisory Committee, (WESIAC), *Final Report of Task Group II, Prediction, Measurement of:* Vol. I, *Summary, Conclusions, Recommendations, AD 458454*, Vol. II, *Concepts, Task Analysis, Principles of Model Construction, AD 458455*, Vol. III, *Technical Supplement, AD 458456*, January 1965.
52. Weapon System Effectiveness Industry Advisory Committee (WESIAC), *Final Report of Task Group IV, Cost Effectiveness Optimization:* Vol. I, *Summary, Conclusions, and Recommendations, AD 458595*, Vol. III, *Technical Supplements, AD 458586,* January 1965.
53. "The Management Decision in Product Strategy and Pricing," Jack Corsiglia, *IEEE Trans. Engineering Management*, **EM-12**, No. 2, 34–43, June 1965.
54. "Reliability and Cost of Avionics," E. J. Nalos and R. B. Schulz, *IEEE Trans. Reliability*, **R-14**, No. 2, 120–130, October 1965.
55. "Outlook on Man's Future," *Technol. Rev.*, M.I.T. Alumni Association, Cambridge, Mass., November 1965.
56. "Annotated Bibliography on Systems Cost Analysis," *Memorandum RM-4848-PR*, P. A. Don Vito, The Rand Corporation, Santa Monica, Calif., February 1966.
57. "The Urban Challenge," *Technol. Rev.*, M.I.T. Alumni Association, Cambridge, Mass., June 1966.
58. "A Methodology for System Engineering: AFSCM 375-5," Norman L. Gelbwaks, IEEE Systems Science and Cybernetics Conference, Washington, D.C., October 1966.
59. "Minuteman System Engineering—A Case Study," Joseph Dresner, IEEE

Systems Science and Cybernetics Conference, Washington, D.C., October 1966.
60. "Economics of Computers in Process Control," T. M. Stout, *Automation,* **13,** No. 11, 82–90, October 1966.
61. "The Systems Man," Norton Gale and Paul Alelyunus, *Space/Aeronautics,* **46,** 81–87, December 1966.
62. *Statistical Models in Engineering,* G. J. Hahn and S. S. Shapiro, Wiley, New York, 1967.
63. "Where the Industries of the Seventies Will Come From," Lawrence Lessing, *Fortune,* **75,** No. 1, 96–99, 184–192, January 1967.
64. "The Road to 1977" by Max Ways, *Fortune,* **75,** No. 1, 93–95, 194–197, January 1967.

Index

The number in parentheses indicates the bibliography number

Abnormal parts, 342
Abnormalities, 341
Acceptance, 39
Acceptance phase, 171, 173
Accomplishment reviews, 43
Accuracy, 89
Achievement, 85
Ackoff, R. L. (9), 70, 73
Acquisition phase, 34
Activities-flow diagram, 75
Activities, project oriented, 20
  technology oriented, 20
Activity, 244
Actual operating environment, 275
Actuating signals, 87
Actuator, 309, 320
Added-value, 167
Added-value classification, 168
Adjacent failures, 310
Affel, H. A., Jr. (47), 2, 40
AFSCM 375-5 (43), 15, 34, 36, 37, 223, 254
Aging, 288
AGREE, 329, 330
Allocation of charges, 179
Alteration notices, 67
Alternative methods, 5, 179
Alternative power sources, 371
Alternatives, 11
Alternative systems, 185
Alternative ways, 359
Applicable documents, 66
Applications, 129
APT, 373
Arnoff, E. L. (9), 70, 73
Arrhenius' relation, 343
ASA C 85.1 - 1963 (39), 89
Audit of reliability, 353
Automated manufacture, 373
Automatic control equipment, 127
Automatic control, functional tools of, 127
Automatic testing, 58
Automatic test system, 58
Automation, 361, 373
Average life, 193
Average-loss rates, 204

Barnes, Wallace, 256
Basic material, 169

Batch-type process, 202
Bath-tub characteristic, 273
Bayesian estimate, 141
Bearings, 319
Bellows, 319
Bennett, Arnold, 221
Bias effects, 328
Blood, Jerome W. (33), 243, 244, 247
Booz, Allen and Hamilton, 243
Bottleneck conditions, 244
Boundaries, 106
Breadboard, 181, 228
Breadboarding, 32
Breadboard model review, 333
Breadboard phase, 13
Break-points, 81
Breakthrough, 83
Brown Boveri Co., 129
Burn-in, 337, 342
Business management systems, 365

Calculating backwards, 235
Calibration, 53
Carnot efficiency, 150
Cash flow, 157
Cause of failure, 340
Centralization, 124
Chain, 247, 258
Chains, 263
Chambers, Dudley, 167
Change, 221
Change memo, 65
Changes, 143
Change with time, 61
Characteristics of the system, 72
Check list, problems, 68
  reliability goals, 352
Check-out, 20, 54, 56, 348
Chemical performance, 148
Chestnut, Harold (15),(31),(49), 2, 5, 13, 18, 73, 121, 146, 171, 201, 217, 236, 241, 317, 359
Choices, 1
Chronological arrangement, 109
Churchman, C. W. (9), 70, 73
Cities, 375
Classes of errors, 94; see also Error
Classification scheme, 44
Clean area, 339
Cleanliness, 337, 338

384 INDEX

Closed loop, 138
Coarse modeling, 110
Coarse regulation, 78
Codes, 89
Coding system, 44
Combinations, 308
Communications, 52, 276
Community development, 374
Community services, 376
Company management, 172
Compatibility, 359
Compatible systems, 185
Completeness, 181
Complexity, 1, 135, 270, 351
Component failure rate data, 302
Component failure rates, 301
Component identification, 44
Component pieces, 169
Components, 300
    conservatively rated, 285
Compound-amount factor, single payment, 159
Compound interest, 159
Computers, 132
Concepts, 360
Conceptual features, 136
Conceptual phase, 34, 171
Confidence level, 279, 282
Configuration inspection, 39
Connecting elements, 171
Consequence of failure, 287
Considerations, customer oriented, 18
Constraint, 210
Constraints, 145
Construction techniques, 373
Consumer, 168
    orientation, 119
Contamination, 338
Contract, 37
Contractor, 173
Control disturbances, 87
Control instrumentation and conditioning, 87
Controlled variables, 87
Control logic, 87
Control performance, 153
Conversion, 148
Coordinate reference systems, 90
Coordinate systems, 93
Coordination, 337, 345
Corrective action, 341
Corsiglia, Jack (53), 156, 157
Cost, 15, 179, 204, 219, 230
    cumulative, 228

Cost-benefit analysis, 210, 219
    reliability of, 216
Cost-benefit studies, 214, 218
    sources of error in, 216
Cost breakdown, 14
Cost comparison, 196
Cost considerations, 187
    medium term, 201
    short term, 204
Cost decisions, long term, 187
    medium-time, 188
    short-time, 188
Cost effectiveness, 14, 210, 219
Cost estimating, 184
Cost expenditures, 228
Cost goals, 190
Cost impact, 213
Costing, 190
Cost of parts, 190
Cost plus fixed fee, 13
Cost-sensitivity analysis, 185
Cost studies, 210
Cost-time trade-offs, 267
Cost versus size-of-equipment, 81
Counting applications, 129
Coupling, 32
CPM, 244, 258
Criterion of value, 10
Critical design review, 39
Criticality of parameters, 111
Critical path, 244, 245, 259, 265
Cross-talk, 32
Custom, 180
Customer, 135, 173, 239
Customers, 168
Customer's objectives, 185

Data feedback, 295
Data nets, 363
Data processing systems, 367
Data storage, 363
Deadband, 89
Debugging, 240
Debugging a new design, 338
Decision-making, 2
Decision-making process, 2
Decision-making rules, 106
Decision rules, 189
Decision tree diagram, 140
Decoupling, 99
Defects, classification of, 341
Defining need, 5
Definition phase, 34
Degradation, 317, 344

Degree of work completeness, 181
Demand, 164
Demand-income, 164
Demand schedule, 187
Demand-supply, 162, 178
Depreciation, 192, 193
Derating of components, 295
Design centered device, 287
Design changes, 65
Design criteria, 27
Design problem checklist, 319
Design review, 30, 172, 288, 293, 331, 333
Design review board, appointment of, 333
Design simplification, 351
Desired values, 137
Developing countries, 377
Development and design phase, 171, 172
Development requirements, 37
Development time, 222
Difference in costs, 212
Differences, 61
Digital computing applications, 130
Dimensionality, 99
Dirt, 338
Discontinuities, 81
Discrepancies, 53
Disturbances, 91, 97, 129
Dobson, D. B. (28), 55
Documentation, 37
Don Vito, P. A. (56), 14, 179
Dresner, Joseph (59), 34
Driving function, 355
Duane, J. T., (44), 213, 232, 233
Dummy, 257, 260
Dynamic characteristics, 90
Dynamic modeling, 267
Dynamic optimization, 237
Dynamic programing, 204
Dynamic response, 91
Dynamic testing, 58

Earliest completion date, 259
Earliest expected time, 246
Echelons of added value, 167
Eckman, D. P. (25), 111
Economic value, 157
EDP, 259
Education, 352
Effectiveness, 210
Efficiency, 87, 146

Elapsed time, 16, 224
Elasticity, 162, 186
Electrical interference, 92
Electric power, 369
End-date requirements, 41
Energy level, 87
Energy systems, 368
Engineering design, 36
Engineering plan, 29
English, J. M. (48), 1, 70, 80
Environment, 11, 75, 274, 289
  considerations, 79
  dynamic conditions of, 291
  mechanical, thermal, 80
  preoperational, 274
  time and money, 76
Environmental factors, 291
Environmental variables, 87
Environments, 76
  organizational, 76, 84
Equipment coding, 45
Equipment design, 183
Equipment lines, 124
Equipment requirements, 115
Equipment structuring, 121, 124
Error, Class A, 94
  Class B, 95
  Class C, 95
  Class D, 95
Essential output, 92
Estimating the time, 16
Event, 244
Evolution, 121
Expected time, 245, 257
Expected time required, 259
Expenditures, 85, 158
Experienced personnel, 353
Experimental test method, 315
Exterior design, 238
External information, 72
Extracted natural resource, 167
Extraneous outputs, 98

Facilities, 184
Factorable systems, 122
Factor of safety, 295
Factorization, 122
Failure consequence study, 317, 322
Failure, definition of, 315
Failure detection, 312
Failure effects, 204
Failure event check list, 320
Failure incidents, 339

Failure mode consequence, 287
Failure mode identification, 318, 322
Failure modes, 315, 322
Failure rate, 273, 277, 286, 300, 338
Failure rate data, 301
Failure-rate modifier, 305
Failure risk, 337
Fatigue-stress, 342
Feasibility, 145
Feasibility estimates, 15
Feedback, 53, 243, 295, 331
Feedback information, 72
Felix, Fremont, (45), 368
Figure of merit, 145
Financial commitments, 181
Fine regulation, 78
Firm-up specifications, 64
First cost, 192
First order effects, 110
Fixed charges, 191, 192, 197
Fixed cost, 13, 206
Fixed-price job, 13, 179
Flagle, C. D. (21), 99
Fleet Ballistic Missile, 267
Flehinger, B. J. (13), 313
Float, 258
Flow diagram, 28
 top-level, 254
Flow of information, 52
Formulating, 70, 104
 the problem, 72
 the system, 18, 72
Formulation, process of, 72
 time, 223
Forrester, J. W. (26), (46), 2, 16,
 119, 268
Fraction of time operating, 206
Frequency, 90
Frequency response, 154, 292
Full-scale system, 183
Functional assembly, 169
Functional element, 169
Functional flow diagram, 254
Functional structuring, 102, 107
Future sum, 159

Gabrielli, G. (2), 152
Gain, 89
Gain coefficients, 327
Gale, Norman (61), 2, 40
Gelbwaks, N. L. (58), 34, 39
Generalized stress, 315
Generic business model, 366
Geographic structure, 102

Gillings, D. W. (18), 190
Goals, 72, 136, 137
Goldman, A. S. (42), 346
Goode, H. H. (16) 238
Gosling, W. (32), 1, 62, 155, 269
Graded program, 349
Graduated tests, 288
Grant, E. L. (23), 158, 161, 190, 191,
 212
Gross return, 209
Grouping of equipment, 130
Growth plans, 189
Growth projections, 189

Hahn, G. J. (62), 163, 214, 216
Hall, A. D. (34), 52, 121, 122, 124,
 164
Handbook of Preferred Circuits, 296,
 297
Hardware, 53
Hatry, Harry, 210
Herd, G. R. (5), 270, 284
Herwald, S. W. (37), 21, 269
Hierarchy, 167
High-rate period, 85
Holahan, James (30), 330
Howard, R. A. (24), 141
Huggins, W. H. (21), 99
Human capabilities, 356
Human engineering, 55
Human environment, 84
Human resources, 61
Human time-constant, 84, 222
Hypo Department, 259
Hypothetical operation, 110
Hysteresis, 89

Ideal operation, 110
Ideal outputs, 93, 94
Identification of equipment, 20
Implementation, 25
Income taxes, 192, 193
Incremental cost, 212
Incremental rate, 82, 209
Incremental utility, 143
Independence, 122
Indexing scheme, 44
Industrial dynamics, 268
Inelastic market, 163
Infant mortality, 273, 340
Inferior good, 164, 177
Influence coefficients, 317
Information collection, 337, 339
Information storage and retrieval 53,
 84

Information systems, 361
Inherent reliability, 336
Initial check-out, 95
Initial formulation, 105
Initial strength, 275
In-operation rate, 30
Input impedance characteristics, 90
Input material, 195, 203
Input-output characteristics, 55
Inputs, 76
   and outputs, 74
Installation, 49, 192, 240
Installed plant capacity, 187
Insurance, 194
Integrating transportation, 377
Interactions, 113
Interconnection of power systems, 368
Interest rate, 159
Interface, 106, 161, 179
Interface problems, 132
Interior design, 238
Intermediate material, 169
Intermediate signal variables, 87
Internal signal inputs, 87
Invariance, 108
Inventory and control systems, 365
Inventory control, 189
Ireson, W. G. (23), 158, 161, 190, 191, 212
Iterative nature, 183
Iterative process, 99, 138

Jervis, E. R. (5), 270, 284
Jet and rocket engines, 369
Job requirements, 180
Job scope, 179
Judgment factors, 2
Jureen, Lars (4), 165

Kettering, C. F., 68
Kirchmayer, L. K. (17), 187
Knight, C. R. (5), 270, 284

Lambert, J. S., 278
Langenwalter, D. F., 126
Large numbers of specimens, 338
Latest allowable time, 246
Latest completion date, 259
Latest possible starting time, 236
Laws of nature, 353
Lessing, Lawrence (63), 355
Level of diagram, 255
Levels of value-added, 167

Life expectation, 193
Life model, 275
Limitations, 28, 110, 355
Limited resources, 202
Linearity, 89
Linearization, 325
Lipp, J. P. (11), 311
Living document, 64
Load applied, 271
Loading, 29
Location structure, 102
Logistics, 58
Logistics system, 211
Long-term equipment evolution, 121
Low-level signals, 133
Luce, R. D. (12), 140
Luxury, 164

Machol, R. E. (16), 238
Magnetic amplifier, 304
Magnitude, 76
Maintenance, 22, 269
   and operating expenses, 192, 194
   environment, 290
   expenses, 212
   for reliability, 346
Major capital expenditure, 191
Major cost components, 191
Malfunction, 340
Management, 36, 253, 269, 345, 349, 352, 353
Management approval, 9
Management systems, 364
Manipulated variables, 206
Man-machine interface, 28
Marginal performance, 174
Marginal costs, 212
Market demand, 186
Material, 192
Material outputs, 97
Material performance, 150
Material systems, 371
Mathematical model, 212
Matrix array, 357
Matrix form, 114
Maximum production, 187
Mayer, R. W. (15), 30, 146, 155
Measure of effectiveness, 211
Mechanical friction, 92
Memory systems, 363
Mesarovic, M. D. (22), (41), 87, 99
Methods, 357
Methods of identification, 53
Miniature region, 81

Minimum time, 225
Minuteman Systems Engineering Plan, 25
Miscellaneous expenses, 192, 194
Miscellaneous loading, 192
Mission objective, 211
Mission reliability, 279
Mod change, 67
Modes of failure, 323, 324
Modes of trouble, 338
Modification changes, 67
Modified material, 169
Modular construction, 373
Modular height units, 47
Modularity, 374
Modular standardization, 58
Modular test, 61
Modules, 60, 373
Monitoring, 242
Monte Carlo analysis, 267
Morton, J. A. (20), 3, 72
Most likely time, 245
Motivation, 350, 351, 352
Motors, 124, 126, 321
MTBF, 330

Nalos, E. J. (54), 144
Natural resource materials, 166, 167
Natural resources, 356
Necessity, 164
Need, 9
Need for system, 5, 7
Net return, 155, 204
Network, 245
Network analysis, 244, 251, 258, 261, 265
    technicalities of, 247
New cities, 375
New equipment, 196
New materials, 372
Nish, J. G. (29), 45, 47
Noise, 91, 97
Nominal performance, 175
Nominal system, 186
Nominal time, 227
Nominal time estimate, 229
Nominal value, 177, 227
Non-dimensional form, 145
Non-linearities, 173
Nonessential outputs, 92
No-repair operation, 347
Norris, R. H., 325, 336, 342, 351
Nozzle actuator system, 309

Objectives, 70, 84, 137, 359
    acceptable set of, 73
    number of, 73
Obsolescence, 241
Old equipment, 195
On-line automatic controls, 189
Open loop, 138
Operating conditions encountered, 274
Operating system, 3
Operating time cycle, 225
Operational phase, 34
Operational requirements, 28
Operational system, 20, 25
Operational system functional analysis, 28
Operation and maintenance time, 223
Optimistic time, 245, 257
Optimization, 359
Optimizing control, 202
Ordered whole, 99
Organization, 20, 24, 62
    project-oriented, 39
    technology-oriented, 39
Organizational environment, 84
Organizational hierarchy, 9
Organizational work groups, 119
Organization interface, 179
Original cost estimate, 184
Other changes, 13, 195
Output variation, 79
Outputs, description of, 96
    direct, 77
    ideal, 93
    indirect, 77
    signal, 96
Overall cost, 191
Overall design, 293
Overall method, 360
Overall system, 36
Overall system requirements, 27
Overhead, 191
Overload, 227
Oversize region, 81
Overtime, 230, 231

Packaging, 290
Parallel failure, 278
Parameter change effects, 328
Parameter changes, 325, 326
Parametric trade-off studies, 172
Part, 169
Partial derivatives, 143, 317
Partitioning, 9, 22

INDEX

Past investment, 13
Past investment and other charges, 191, 195, 199
Path, 258
Path float, 258, 263, 264
Peak resources, 236
Penalty clauses, 176
Penalty costs, 203
PEP, 244
Performance, 145, 229, 325
Performance capability, 241
Performance characteristics, 131
Performance degradation, 316
Performance indices, 146
Performance needs, 284
Performance objective, 144
Performance requirements, 32, 34
Periodic checks, 242
Periodic maintenance, 189
PERT, 14, 43, 222, 243, 247, 255, 266
PERT/COST, 244, 267
Per unit system, 147
Pessimistic time, 245, 257
Physical environment, 75, 77
Physical needs, 110
Piece, 169
Planning review, 332
Polaris, 243
Pollution-control systems, 375
Pontryagin's maximum principle, 236
Poorly defined problems, 5
Potential need, 8
Power, 79
Power engineering application, 129
Power inputs, 87
Power level, 91, 129
Power outputs, 97
Power requirement, 79
Power source identification, 90
Power supplies, 132
Power supply variations, 326
Predecessor event, 261
Prediction, 202
Predictive control, 243
Preferred circuits, 296
Preferred system, 210, 212
Preliminary design, 183, 293
Preliminary design requirements, 27
Preliminary design review, 37
Preliminary specifications, 74
Preoperational environment, 274, 289
Preparation for delivery, 66
Present-worth concept, 158, 160

Present-worth factor, single payment, 160
Preventive maintenance, 347
Price level changes, 196
Pricing, 82
Primary end-use material, 169
Primary power, 91
Primary system characteristics, 212
Priority, 231
Probability, 258, 270, 271, 277
Probability density, 314
Probability distribution, 272
Probability formulas, 279
Probability of success, 140, 246
Problem checklist, 318
Problem definition studies, 15
Problem formulation, 98
Problem objectives, 72
Problem structuring, 98
Processing of final output, 195
Process variables, 206
Product, 180
Product demand curve, 202
Product focus, 119
Production, 181
Production efficiency, 206
Production phase, 13
Production release review, 333
Production time, 222
Production volume, 337
Product or service, cost of, 155
  value of, 155
Products and services, 168
Programing, 189
Program orientation, 40
Progressive factorization, 122
Progressive systematization, 122, 123
Project, 120
Project engineer, 346
Property taxes, 192, 194
Proposal phase, 171
Proposal review, 332
Prototype, 40, 136, 181, 228
Prototype design review, 333
Prototype phase, 13,
Pumps, 321

Quality assurance, 66
Quality control, 34, 63, 275, 290, 346

Raiffa, Howard (12), 140
Random effects, 328
Random failure, 300, 340
Random-failure method, 300

Real time, 363
Real world, 94, 355
Rebuilding city areas, 374
Reconciliation, 242
Record-keeping, 24, 52, 53
Record-keeping procedure, 20
Redundancy, 278, 307, 311
Reference, 136, 137
Refined natural resource, 169
Refrigerators, 124, 126
Regulation, 91
Regulation, coarse, 78
    fine, 78
Regulation of physical environment, 77
Relative luxuries, 165
Relative size, 76
Relative weighting, 73, 143
Relative worth, 142
Reliability, 21, 22, 232, 269, 270
    analysis, 323
    approach to, 270
    arithmetic, 277
    block diagram, 283
    check list, 351
    current developments, 270
    demonstrations, 329
    designing for, 284
    enhancement, 308, 311
    goals, 331, 352
    monitoring, 278
    requirements, 350
Remote access, 362
Repetition frequency, 131
Repetition rate, 90
Requirements, 37, 66
Resources, 155, 236
Restructuring, 105
Return on investment, 192
Review, 41
Review design, 288, 331
Review reports, 335
RFP, 183
Risk, 141, 253
Risk analysis, 352
Risk assessment, 140
Roy, R. H. (21), 99
Ruggedness, 130
Run-in, 342, 344

Salvage, 193
Salvage value, 196
Sandler, G. H. (38), 296, 277
Saturation, 89

Schedule, 41, 243
Schedule preparation, 242
Scheduling, 24, 41, 189
Scheduling and review procedure, 20
Schematic block diagram, 29
Schematic drawings, 53
Schroeder, T. W. (19), 191, 196
Schulz, R. B. (54), 144
Scientific orientation, 120
Scope, 66
Scrap material, 195
Screening methods, 337, 342
Sensitivity, 89, 233
Sensitivity analysis, 111, 216, 217
Series failure, 277
Service, 180
Serviceability, 32
Servicing, 49, 54
Shapiro, S. S. (62), 163, 214, 216
Ship, installation, and check-out time, 223
Shipment of a device, 290
Short term, 204
Shut-down costs, 203
Signal disturbances, 87
Signal equivalency, 90
Signal inputs, 87
Signal outputs, 96
Signal source identification, 89
Signal-to-noise ratio, 92
Significance, 135
Similarities, 5, 61
Similarities and/or differences, 61, 109
Simplification of design, 351
Simulated conditions, 136
Single payment, 159, 161
SKETCHPAD, 374
Slack, 226, 246
Slattery, T. B. (42), 346
Software, 36
Space, 356
Space systems, 211
Specification format, 65
Specifications, 18, 62, 66, 184
    changes to, 64
    early, 63
Specification writing, 74
Spread, 245
Stages of added value, 167
Standard building blocks, 60
Standardization, 47, 82, 300
    lack of, 299
Standardized chassis construction, 45

Standard signals, 131
Standby costs, 203
Starting period, 85
Start-up costs, 203
State of the art, 80, 175, 241
State of the equipment environment, 75
Static testing, 58
Statistical methods, 288
Statistical treatment, 218
Stimuli, 58
Stopping period, 85
Storage, 290
Storage and retrieval systems, 84, 363
Stout, T. M. (60), 12, 155
Strength, 271
    and stress, 286, 313
    versus time, 273
Stress, 272
Stress requirements, 284, 289
Stress-strength relationship, 286
Structure, 18, 98
Structuring, 70, 99, 104
    concepts, 107
    by equipment, 113, 124
    equipment lines, 124
    functional, 107
    the problem, 72
    of a system, 18, 70
    by time phases, 100, 114
Study, 181
    contracts, 15
    phase, 13
Subdivide, 190
Subdivision of costs, 196
Suboptimizations, 213
Subsystem, 93
Successive approximations, 99
Successor event, 262
Superclean area, 339
Support data, 28
Support system, 20, 25
    evaluation, 30
    functional analysis, 29
    requirements, 30
    technical requirements, 29
Switching, 312
System, 1, 36, 357
    concepts, 27, 65
    configuration, 172
    cost, 12, 201
    designer, 87
    design evaluation, 32
    development, 143
    documentation, 37

System, environment, 75
    evaluation, 29, 40
    function requirements, 116
    inputs, 87
    installed configuration, 29
    investment, 192
    judging the value of, 135
    life cycle, 34
    objectives, 74
    operational phases, 118
    outputs, 92
    process of making, 5
    redefinition, 8
    studies, 191
    suppliers, 6
    time phase, 171
    tree, 49
    users, 6
    value, 173
    work elements, 5
System engineering, management, 36
    methods, 1
    plan, 25
Systemization, 122, 123
System requirements, 136, 142
    establishing the, 137
Systems, approach, 360
    opportunities, 361
    planning, 189
    requirements, 20, 142
    structuring, drawbacks of, 106
Systems engineering, 36, 99, 355, 361
    approach, 360
    definition of, 3
    methods, 2
    organization, 39
    overall problem, 3
    process, 30
    tools, 2, 22, 73, 236
Systems engineers, 41

TANES, 243, 255, 265, 267
Task identification, 261
Task network scheduling, 255
Taxes, 192
Technical development plan, 34
Technical requirements, 29
Technical risk, 36
Technology orientation, 40
Telemetry, 129
Telephone network, long distance, 123
Temperature cycling, 344
Ten-best-problems list, 68

Tentative designs, 285
Tentative system specifications, 74
Terminal time, 225, 236
Terminal value problem, 222
Test data, 367
Test equipment, 55
Test equipment, general purpose, 60
  special purpose, 60
Testing, 54, 287
Testing severity, 348
Test performance, 53
Tests, 287, 325, 329
Test types, 57
Thermal, 149
Threshold, 91
Time, 15, 221, 222, 227
  boundaries, 233
  constant, human, 84, 222
  to develop and design, 239
  estimates, 245
  estimation, 16
  to formulate, 239
  history, 156
  to make system, 237
  and money, 76
  and money environment, 85
  phase, 171, 190
  to produce and test, 240
  schedules, 241
  sequence of events, 254
  to ship and install, 240
  structure, 100, 241
  for system to operate, 240
Time-line analysis, 29, 30
Time-shared, 362
Tolerances, 287, 325
Tools, 357
Top-management support, 350
Tornqvist, 165
Total cost, 13, 155, 183, 212
Total cost, buildup of, 190
Total costs, range of, 183
Total elapsed time, 224
Total incremental utility, 143
Total requirements, 25
Total system, 121
Total time, 227
Trade-off, 174, 178, 179, 210
  knowledge, 74
  studies, 29
Training and maintenance, 184
Training of personnel, 36
Transfer functions, 55

Transistor preferred circuits, 298
Transportation, 376, 377
  performance, 152
  systems, 376
Transport efficiency, 152
Triode drifts, 326
Tune-up, 240

Ultra-clean room, 339
Unbalance forces, 92
Uncertainty, 180, 216
Unit minutes of testing, 283
Unregulated environment, 78
Utility, 143

Validation, 36
Value, 9, 12, 135, 145, 155, 161, 204
Value-added, 166
  classification, 168
Value-cost, 176
Value effectiveness, 219
Value objectives, 10
Value of money, 160
Value-performance, 174
Value-reliability, 177
Value-time, 175
Variable charges, 191, 194
Variable cost, 13, 198, 201, 202
Variation of inputs, 78
Variations, 325
Variations in power sources, 92
Vendor, 135
Voltage stimulus, 56
Von Karman, T. J. (2), 152

Waste material, 195
Wave-shape types, 55
Ways, Max (64), 9, 355
Wear-out, 273, 340
Weighting factor, 143
WESIAC (50), (51), (52), 14, 135, 140, 210, 219
Whole system, 121
Wilcox, R. B. (35), 16, 221, 268
Wold, Herman (4), 165
Wolff, L. L. (28), 55
Work functions, 2
Worth, 2, 135

Yield, 148